Climate Change and Justice

Achieving climate justice is increasingly recognised as one of the key problems associated with climate change. The numerous and complex issues which climate change involves underline the need for a normative framework that allows us both to assess the dangers that we face and to create a just distribution of the costs of action. This collection of original essays by leading scholars sheds new light on the key problems of climate justice, offering innovative treatments of a range of issues including international environmental institutions, geoengineering, carbon budgets and impact on future generations. It will be a valuable resource for researchers and upper-level students of ethics, environmental studies and political philosophy.

JEREMY MOSS is Professor of Political Philosophy at the University of New South Wales, Sydney. His publications include *Reassessing Egalitarianism* (2014).

Climate Change and Justice

Edited by
Jeremy Moss

CAMBRIDGE
UNIVERSITY PRESS

University Printing House, Cambridge CB2 8BS, United Kingdom

Cambridge University Press is part of the University of Cambridge.

It furthers the University's mission by disseminating knowledge in the pursuit of education, learning and research at the highest international levels of excellence.

www.cambridge.org
Information on this title: www.cambridge.org/9781107093751

© Cambridge University Press 2015

This publication is in copyright. Subject to statutory exception and to the provisions of relevant collective licensing agreements, no reproduction of any part may take place without the written permission of Cambridge University Press.

First published 2015

A catalogue record for this publication is available from the British Library

Library of Congress Cataloguing in Publication data
Climate change and justice / edited by Jeremy Moss.
  pages cm
Includes bibliographical references and index.
ISBN 978-1-107-09375-1
1. Environmental justice. 2. Climatic changes – Social aspects.
3. Global environmental change – Social aspects. I. Moss, Jeremy.
GE220.C55   2015
363.738'74–dc23
                                                            2015018508

ISBN 978-1-107-09375-1 Hardback

Cambridge University Press has no responsibility for the persistence or accuracy of URLs for external or third-party internet websites referred to in this publication, and does not guarantee that any content on such websites is, or will remain, accurate or appropriate.

Contents

List of contributors		*page* vii
Acknowledgements		viii
	Introduction: climate justice JEREMY MOSS	1
1	Climate change and state consent THOMAS CHRISTIANO	17
2	Geoengineering in a climate of uncertainty MEGAN BLOMFIELD	39
3	Climate justice and territorial rights CHRIS ARMSTRONG	59
4	Exporting harm JEREMY MOSS	73
5	What's wrong with trading emission rights? AXEL GOSSERIES	89
6	A just distribution of climate burdens and benefits: a luck egalitarian view KASPER LIPPERT-RASMUSSEN	107
7	Individual duties of climate justice under non-ideal conditions KOK-CHOR TAN	129
8	Acts, omissions, emissions GARRETT CULLITY	148
9	Individual responsibility for carbon emissions: is there anything wrong with overdetermining harm? CHRISTIAN BARRY AND GERHARD ØVERLAND	165
10	Climate change: life and death JOHN BROOME	184

11	What we have done ≠ what they can do	201
	BENJAMIN HALE	
12	Empathising with scepticism about climate change	219
	SIMON KELLER	

Bibliography 236
Index 249

Contributors

Chris Armstrong, University of Southampton

Christian Barry, Australian National University

Megan Blomfield, University of Bristol

John Broome, University of Oxford

Thomas Christiano, University of Arizona

Garrett Cullity, University of Adelaide

Axel Gosseries, University of Louvain

Benajamin Hale, University of Colorado, Boulder

Simon Keller, Victoria University of Wellington

Kasper Lippert-Rasmussen, University of Århus and University of Tromso

Jeremy Moss, University of New South Wales

Gerhard Øverland, University of Oslo

Kok-Chor Tan, University of Pennsylvania

Acknowledgements

This book began as an idea for a collection on climate change with my colleagues Simon Keller, Iwao Hirose and Garrett Cullity. I would like to thank them all for the effort they have, at various times, put into getting the book organised, especially the workshops where some of the papers were discussed. Garrett in particular has been very generous with his time and support for the collection. The book has had the crucial support of a number of grants from the Australian Research Council (ARC) – they enabled me to have time off for research and travel to other universities – in particular an ARC Future Fellowship and a Discovery grant, on Climate Justice. I thank the Council for its generous support. I would also like to thank Alicia Coram for crucial proof reading and formatting assistance.

Introduction: climate justice

Jeremy Moss

It has been well documented that the kind of dangerous climate change with which we are faced is likely to produce major harms for the planet and those who live on it. The increased spread of disease, incidence of extreme weather events, sea-level rise, disruption to agriculture and so on are just some of the very significant harms that are likely to be caused by anthropogenic climate change. Ethics plays a crucial role in our understanding of the problem of climate change through determining how good or bad the effects of climate change are, identifying harms, establishing how severe these harms are and by comparing them to other harms. A related question to how we evaluate the harms of climate change is how we fairly distribute the right to emit, trade, offset and measure greenhouse gases (GHGs). One of the reasons that these issues are important is because of the potential for causing further harm by imposing unfair distributive arrangements as part of any climate agreement. A distribution that placed a large part of the burden of climate mitigation on the already disadvantaged, for instance, would increase the impact of harms and threaten to further exacerbate their disadvantage. Settling these questions involves engagement with broader theories of justice especially distributive principles, the procedures that accompany them, as well as the broader goals of climate justice.

These two aspects of climate change justice – consideration of the kinds of harms and the justice or otherwise of how those harms are shared – underscore why we need a normative framework that allows us to assess the dangers that we face as well as one that allows us to create a just distribution of the costs of action to prevent dangerous climate change. This is a challenge to which moral and political philosophy has a great deal to contribute. The chapters in this collection tackle many of these issues in new and interesting ways. Before discussing them, I will provide a brief discussion of some of the issues that frame an account of climate justice. In particular, how we ought to view the kinds of principles that ought to regulate the distribution of rights to emit greenhouse gases, which in many ways frames the current discussion of climate justice.

Allocating burdens

The burden created by the threat of dangerous climate change can take different forms ranging from the division of the remaining emissions, the costs of mitigation and adaptation including opportunities foregone, and compensating those who have been wrongly harmed. The most contentious issue for the climate debate to date has been how to allocate the remaining emission rights. At issue is what principle or combination of distributive principles should be used to decide on how to divide the costs of action to prevent dangerous climate change. Along with the science concerning the likely effects of different degrees of warming, these justice-based principles would ideally form part of the framework for any distribution of climate costs. At work in the debate are several different kinds of distributive principles: fault-based, benefit, equality and ability principles. Fault-based principles, often called 'historical responsibility', 'polluter pays', 'harm', 'contribution' or simply 'fairness' principles, require that the costs of action to mitigate or adapt to climate change should fall proportionally on those who have played the greatest role in contributing to those harms or risk of harms. What each version shares is the thought that there is a causal link between past actions that have contributed to some kind of harm and the liability to bear some of the costs of that harm.[1] In terms of the 'carbon budget' debate, the claim is that those countries who are causing or have caused a harm have a prima facie reason to shoulder the cost of addressing the effects of that harm, or, in the case of claims to emit in the future, to have their current or past emissions counted against any fair distribution.[2] One reason why this approach is relevant is because the total anthropogenic emissions have largely been emitted before the science of climate change was widely known.[3] While fault-based principles also apply to emissions that have been emitted in the very recent past and for those that are being emitted now, for convenience I will use historical emissions primarily to refer to emissions prior to the widespread dissemination of scientific knowledge of climate change.

The most drastic outcome of this position is that states guilty of large historical emissions would assume strict liability for all the costs associated with the harm to which they have contributed. One of the difficulties for fault-based principles that argue for strict liability is that often the form of the

---

[1] For discussion see: A. Gosseries, 'Historical emissions and free-riding', *Ethical Perspectives*, 11/1 (2004), 36–60; L. H Meyer, 'Compensating wrongless historical emissions of greenhouse gases', *Ethical Perspectives*, 11/1 (2004), 20–35; E. Neumayer, 'In defence of historical accountability for greenhouse gas emissions', *Ecological Economics*, 33 (2000), 185–92.
[2] The carbon budget is the amount of GHGs that we can emit before we risk dangerous climate change and we have already emitted more than half of it. M. Meinshausen *et al.* 'Greenhouse-gas emission targets for limiting global warming to 2 degrees C', *Nature*, 458 (2009), 1158–62.
[3] A. Grubler and N. Nakicenovic, 'International burden sharing in greenhouse gas reduction', *Institute for Applied Systems Analysis* (Laxenburg, Austria, June 1994), pp. 15–16.

principle is best suited to individuals and not parties. While it is easier to assign liability to a person living today who made a decision in the past as they are the same person, the situation of states is different. The question is whether states should inherit the consequences of actions of those in the distant past. What we might call 'weak link' problems beset applying fault-based principles to states and historical emissions. For strict liability to apply we ought to be confident that the state has the characteristics that allow duties to be transmitted where current and future members can be liable for the actions of past members. There are several versions of such problems. There are problems of 'broken transmission'. For instance, given that in many cases those who caused emissions in the past are long dead, we need a further argument concerning how the costs of their actions ought to be assigned. Second, there may be problems of what we might call 'legitimate repudiation'. For example, where the elite of a previous generation took on exorbitant amounts of debt to fund a lifestyle available only to them, the next generation has a plausible case for claiming that they are not the ones who ought to pay for this debt. In the case of climate change, current generations may argue that they should not have to sacrifice now as a result of the emissions debts of their forebears where these have not produced a lasting or well-distributed gain. A third issue arises when some in the state may have good reasons not to be bound by the state's actions. If an individual did not consent to the polluting policies and made significant efforts to mitigate his own behaviour (and perhaps the behaviour of others), then there may be 'exonerating circumstances'. Given that the bulk of the emissions that have caused today's climate change were emitted before people could reasonably be expected to have known about the effects of what they were doing, it is at least plausible to argue that responsibility might be reduced or even cancelled where there was 'reasonable ignorance'. All these kinds of cases raise questions about whether the link between the state and the citizen is strong enough to translate a state's past contributions to harms into current duties to ameliorate those harms.

Beneficiary principle

The contrast between the quality of life of a person in a developed country and that of someone in an undeveloped country is large, and caused in part by greater economic development with its associated GHG emissions. The fact that some countries have benefitted – even unwittingly – from this industrialisation and its associated emissions is an alternative source of duties to apportion some of the costs of climate change. The beneficiary principle, as it is often called, has different forms and applications. But the general idea is that if an agent benefits from an injustice that also causes harm to another agent or agents, the agent who has benefitted has a duty to compensate those harmed to the value of the benefit gained. Benefitting in this sense is different from cases where the beneficiary is an accessory to a harm, that is, where they are involved in the actions that helped bring about a harm.[4] In the case of climate change (and taking countries as the relevant actors), the beneficiary principle has an obvious application in relation to understanding the generation of duties to combat climate change. Countries whose recent prosperity is built on the benefits of emission intensive industrialisation have benefitted unjustly from access to unrestricted emissions even though they may have been ignorant of the effects. Unlike fault-based principles, the beneficiary principle standardly refers to cases where the beneficiary is not part of the causal chain that caused the harm in the first place. Moreover, the benefits here in question are more than the fruits of a lucky error in someone's favour. It is central to the principle that the benefits accrue innocently (the beneficiary did not cause the harm) but also as a result of an injustice or wrongdoing.

Whether we find the use of the beneficiary principle more or less compelling will depend in part on how much of a role it plays in allocating duties to pay the costs of climate change. The strongest use of the principle would be to say that it was the only principle that mattered for the distribution of costs. Yet most of its proponents use the principle in conjunction with fault-based or ability-based principles. Caney, for instance, uses benefitting to augment the ability principle, employing it to help cover the allocation of those emissions that the ability principle doesn't address (what he usefully calls the 'Remainder').[5] This seems also to be the most common application of the principle as part of a broader approach, which can include fault-based principles.

Nonetheless, even where the beneficiary principle is combined with other principles there are issues that must be addressed. Leaving practical issues aside, the problems that persist for many different versions of the beneficiary pays principle concern to what I call 'proportionality'. Let me give some examples.

The first concerns what to say in relation to countries that have accrued a benefit as a result of emissions but since lost that benefit or a proportion of it. For example, take the problem of what Ed Page calls 'Rich Then, Poor Now', where a state has benefitted from the unjust use of emissions but has since squandered or lost the benefit.[6] In a variant of this problem, a country may have benefitted but those benefits went to only a few, as can occur in countries afflicted by the so called 'resource curse'. Second, imagine that a country had

---

[4] E. Page, 'Give it up for climate change: a defence of the beneficiary pays principle', *International Theory*, 4 (2012), 300–30; R. Goodin and C. Barry, 'Benefitting from the wrong doing of others', *Journal of Applied Philosophy*, 31/2 (2014), 363–76; Gosseries, 'Historical emissions and free-riding'.

[5] S. Caney, 'Climate change and the duties of the advantaged', *Critical Review of International Social and Political Philosophy*, 13/1 (2010), 203–28.

[6] E. Page, 'Climatic justice and the fair distribution of atmospheric burdens: a conjunctive account', *Monist*, 94/3 (2011), 412–32.

benefitted enormously from its emissions intensive industrialisation. Suppose further that there were several other countries in a similar situation. If the climate harms that have resulted from this benefit only cost a proportion of the benefit to fix, say 50 per cent, what should be done with the remainder of the benefit? Some argue that the beneficiary principle requires the beneficiary to relinquish all of the benefit where it was gained unjustly.[7] Others argue that this needs further argument. A similar and more likely scenario is that the costs of the harm exceed the benefit.

Ability to pay

The ability to pay principle is often invoked as a way of dealing with those emissions that cannot be easily accounted for, such as distant historical emissions from the long dead, without inflicting major sacrifices on those who cannot afford it. Roughly stated the principle is that those who are able to alleviate or mitigate harm ought to do so, even if they are not themselves responsible for that harm. In the case of climate change, this may mean that wealthy states would pay (or pay more of) the costs of climate mitigation and perhaps adaptation. Its appeal is that it allocates costs to those who can most afford to bear them, and spares burdening the poor with a cost that they would struggle to afford.[8] The rate at which countries pay can be proportional in that the super wealthy pay more than the merely wealthy and so on.

However, the appeal of this principle is likely to be less where it is the sole principle that regulates the distribution of emissions rights or mitigation costs. For instance, one objection is that the ability to pay principle does not take into account morally relevant considerations such as those causal links represented by fault-based principles. This seems true in the case of current or very recent emissions, but less so for those emissions from the distant past. If the ability to pay principle was to count as the only principle this would be a serious objection.[9] Similarly, some might argue that the wealthy are simply entitled to their wealth and the fact that this means the poor have extra burdens is not an injustice, though it might be bad in obvious ways. A response to this objection is to ask what would be the case if one were not to adopt an ability to pay principle in this context. If the wealthy were to assert that it is not their fault and

---

[7] Goodin and Barry, 'Benefitting from the wrong doing of others'. For a contrary view see D. Butt, 'A doctrine quite new and altogether untenable: defending the beneficiary pays principle', *Journal of Applied Philosophy*, 31/4 (2014), 336–48.

[8] For discussions see D. Miller, 'Global justice and climate change: how responsibilities should be distributed', *The Tanner Lectures on Human Values*, Tsinghua University, Beijing, 24–5 March (2008); H. Shue, 'Global environment and international inequality', *International Affairs*, 75/3 (2003), 531–45. Peter Singer's approach in P. Singer, 'Famine, affluence and morality', *Philosophy and Public Affairs*, 1/1 (1972) 229–43 is also closely related.

[9] See Caney 'Climate change and the duties of the advantaged'.

hence they should not pay, then the same is true for the poor, and they will be seriously disadvantaged by having to bear the burden. Unless all parties decide that business as usual is the best option, then someone will have to pay for mitigation and adaptation costs and it seems unfair if the poor assume that burden instead of the rich.

A second response to the objection that we should not ignore fault is to combine the ability to pay with some consideration of how wealth was gained. If, for instance, the wealthy have gained their wealth in ways that might diminish their entitlement, say in ways that are no longer acceptable such as colonialism, then their claim of entitlement is plausibly weaker. So too where the wealth has been gained through imposing externalities on others even if they are innocent, as is the case with dirty industrialisation.[10] In each case the claim to entitlement is less than it would otherwise be. This strategy does raise the question of how much the ability to pay principle overlaps with either the benefit or fault principles. If historical considerations alter or determine how wealth is allocated by, in this case, inquiring how the wealth was caused, then this seems to strengthen the scope and relevance of backward looking principles in climate burden allocations.

Equal per capita

One suggestion for how to avoid at least some of the uncertainty associated with the three principles above is to limit the contribution that a country can make via a fixed quantity principle, such as strict equality. Along these lines the Equal Per Capita (EPC) approach has been suggested, especially in relation to emissions budget, as one way of distributing burdens. While it is not a standalone principle, it draws on egalitarianism and has been widely supported.[11] The EPC approach simply divides the amount of GHGs that we can emit by the number of people in the world (usually adjusted to reflect a population baseline year), which gives us a figure of how many tonnes of GHG equivalent gases we could each emit as a baseline for allocating emissions entitlements. The justification for the EPC appeals to some fundamental intuitions. For instance, Baer argues that everyone has a prima facie claim to a share of the Earth's resources because it is a kind of global commons.[12] As Singer writes, 'Why

---

[10] See *op. cit.*
[11] P. Baer, 'Equity, greenhouse gas emissions, and global common resources', in S. Schneider *et al.* (eds.), *Climate Change Policy: a Survey* (Washington DC: Island Press, 2002), p. 401. See also: T. Athanasiou and P. Baer, *Dead Heat: Global Justice and Global Warming* (New York: Seven Stories Press, 2002), p. 28. A. Meyer, *Contraction and Convergence: the Global Solution to Climate Change* (Dartington, UK: Green Books, 2000); J. Moss, *Reassessing Egalitarianism* (London: Palgrave McMillan, 2014), ch. 4.
[12] Baer, 'Equity, greenhouse gas emissions, and global common resources', p. 401. See also Athanasiou and Baer, *Dead Heat: Global Justice and Global Warming*.

should anyone have a greater claim to part of the atmospheric sink than any other?'[13] While departures from EPC could be justified, each person has an equal prima facie claim to available world resources.[14] Others such as Pogge claim that people maintain a 'minority stake' in global resources.[15]

Yet, the difficulties for the EPC approach also stem from its lack of flexibility and scope. As Caney and Bell have noted, it fixates on one small dimension of environmental justice – the ability to emit – and applies the principle of equality to this dimension alone, whereas what we should be interested in is whether we have equal access to a whole package of goods.[16] For instance, just as we do not have a separate distributive principle for the allocation of iron ore or rare earths, we should not have a separate one for emissions. Emissions gain their importance from the contribution they make to other goods, and it is those goods to which we should apply a principle of equality. This is what we might call the *scope* objection. In addition, if we consider mitigation more broadly as both reducing GHG emissions and utilising GHG sinks, then it would seem we should have a distributive principle that encompasses both of these elements and not just emissions. Similarly, if person A emits the same amount as person B, but they also promote or contribute to the transfer of green energy technology, then this contribution to mitigation should also be part of the equation. This we might call the *incomplete response* objection.

Each of the approaches to how we ought to divide the costs of responding to climate change faces difficulties. Some of those difficulties can be mitigated when the principles are combined. This seems necessary in order to capture the complexity of the situation where we have past emitters, current beneficiaries, or parties as well as clearly culpable current emitters and non-climate related factors such as extreme poverty.

A just procedure

While getting the right principles to regulate the distribution of goods such as emission rights is crucial, so too is the correct procedure for arriving at the principles. There are two broad methods for arriving at treaties. The conventional wisdom is that only a process that is fully inclusive and fair to all parties will ensure that a treaty on climate change has legitimacy. Therefore states and the people whom they represent have to be involved in deciding these issues. This 'universal method' of treaty construction is exemplified in one way by the United Nations Framework Convention on Climate Change

---

[13] P. Singer, *One World* (Melbourne: Text Publishing, 2002), p. 39.
[14] C. Beitz *Political Theory and International Relations* (Princeton University Press 1979), p. 141.
[15] T. Pogge, 'An egalitarian law of peoples', *Philosophy and Public Affairs*, 23/3 (1994), 195–224, p. 200.
[16] S. Caney, 'Just emissions', *Philosophy and Public Affairs*, 40/4 (2013), 255–300.

(UNFCCC). Its characteristics are near universal agreement among parties, a low level of demands (with the possibility of increased demands in the future), a consensus-based approach, weak enforcement of agreements and a process that attempts to deepen the commitments of parties as it proceeds. In contrast to the universal approach the 'club method' begins with relatively deep commitments among fewer states and then expands the treaty by allowing other states to join. A treaty such as the General Agreement on Tariffs and Trade (GATT) is an example of this model.

Thomas Christiano's chapter questions this endorsement of the universal method. Christiano notes that for a process of creating international law to be legitimate, it must involve state consent, but that when the agreement concerns the pursuit of morally mandatory aims such as alleviating global poverty or climate change, there must be further constraints on the reasons given for withholding consent. Where states employ reasons that are, for example, irrational, unscrupulous or morally self-defeating in not consenting to the pursuit of morally mandatory aims, it is acceptable for other states to pressure them to do so. Thus the club method may offer acceptable advantages over the universal method (which to date has not been very effective), and opens up possibilities such as multi-state agreements on GHG reductions. One problem for this method is that the states that are part of the initial agreement necessarily have more say in the details of the agreement. Christiano notes that this leads to a kind of path dependence. Yet one response that he develops in the chapter is that this does not necessarily violate the demand of having an equal say in an agreement. Parties to an agreement may have different levels of input depending on what is at stake for them.

Establishing the right process whereby decisions are made on measures to combat climate change is also important in relation to specific measures. For example, procedural issues are central for any proposal for geoengineering the planet through solar radiation management, via cloud whitening or injecting particles into the stratosphere, and carbon dioxide removal through enhanced carbon sinks. There are many reasons why people have raised concerns about geoengineering, including that it is 'playing God' with the planet, or that it derails other attempts at mitigation such as the transition to renewable energy. Yet as Megan Blomfield points out, certain kinds of geoengineering techniques such as ocean fertilisation or solar radiation management are of concern because of their potential global effects, their uneven distribution and the general uncertainty surrounding their impact. She argues that such features demand a 'governance first' rather than a research first approach.

Blomfield claims that the current uncertainty that surrounds geoengineering could even have some normative value. Just as Rawls's original position was supposed to produce a fairer outcome because the parties would be uncertain about how the costs and benefits of various rules would affect them, so too with uncertainty regarding geoengineering. Parties may be more likely to agree on fair rules for how to govern research where they are uncertain on their position in the distribution of costs and benefits. The uncertainty that we currently have regarding geoengineering might then help produce a fairer agreement. Given the uncertainty parties may seek to design governance mechanisms that, for instance, offer special protections for the disadvantaged.

Carbon budgets

We saw above that the EPC approach to dividing the world's remaining carbon budget draws part of its appeal from the idea that no one has any greater claim to the atmosphere's sequestering capacity. A common response to this position is to claim that it does not make sense to insist on equal shares of this good in particular, and that we should also consider a country's carbon budget in light of their efforts to reduce their emissions in other ways, such as developing renewable energy technologies. However, Chris Armstrong argues that there remain questions over ownership rights of terrestrial sinks like forests that may undermine the justification for the EPC position itself. Armstrong considers, and ultimately rejects, three bases for claims of ownership by particular groups. The first is attachment. A community who controls a rainforest may be very attached to it in the sense that their identities are bound up with access to it. Such a reason to exercise control over a terrestrial sink may not trump all other reasons or even equate to full control, but it might provide a *pro tanto* reason to prioritise their claims. Similarly, communities may claim that access to particular sinks is necessary for self-determination. Again this claim would need to be qualified according to how much of the resource and what degree of control is required to enable self-determination. The third type of claim stems from a party's efforts to improve a sink, which might draw on desert or responsibility-catering arguments. However, Armstrong claims that a better basis for the argument that states ought to get credit in some cases for terrestrial sinks is the sacrifices that are incurred by keeping sinks as opposed to using them for other purposes related to economic development.

The question of how to allocate the costs of climate change and in particular the remaining carbon budget is also complicated by another kind of issue. States are said to be responsible for the emissions they produce within their own territory and by extension for any harms that they may cause. However, the emissions that are produced from products that they export, such as coal, are the responsibility of the importing states. The same formula applies to determining a state's carbon budget. Yet there is an argument to suggest that this straightforward formula for allocating responsibility is too simple. For instance, consider other commodities such as uranium, tobacco or medical waste. In the case of uranium, if one state exported it to another state knowing full well that the state was unstable and likely not to keep the fuel from falling into the hands of rogue military groups, or failed to maintain adequate safety standards, we would apportion some of the responsibility for the resulting harms to the exporting state. What all of the cases above have in common is that they involve goods that cause harm in a morally significant and blameworthy way. Similarly, Moss argues that we should apply the harm principle to the export of commodities such as coal, gas or other fossil fuels. If this is the case, then countries ought to assume some responsibility for the harm caused by their exports of fossil fuels.

If these claims that harms from fossil fuel emissions are the moral responsibility of exporting countries are correct, it has potentially dramatic consequences for how we apportion the costs of climate change as well as how we determine a country's carbon budget. While exporting countries may not have full responsibility for the harms that their exports cause, they plausibly have some. Similarly, accepting responsibility in this way may lead to a revision of a state's carbon budget.

Carbon trading

The discussion of whether there might be parallels between the responsibility that states accept in relation to commodities such as medical waste and their responsibility for fossil fuels also raises the question of whether we should permit the trading of emission quotas at all. As Axel Gosseries notes, where this trading is tied to a cap that keeps the world on track to avoid dangerous climate change, it has several potential advantages. First, it offers a degree of flexibility in how countries can respond to climate change. This will be advantageous where one country has difficulty decarbonising its economy or requires more time to do so. Trading also potentially allows emissions reductions to occur where they are cheapest, creating efficiency. Emissions trading also allows the creation of a market that provides information on how hard it is to meet reduction goals which can be used by regulators to adjust the cap if need be.

However, critics object to emissions trading on several grounds. One of the best known objections is that it grants states the legal right to do wrong. In response, Gosseries argues that buyers of permits are buying the right to do something that is not wrongful because of the corresponding duty to reduce emissions by an equivalent amount that others agree to. Gosseries considers four other objections to trading emission rights, including: whether sellers should get paid for what they should do anyway, whether buyers and sellers lose track of the wrongness of emissions when trading, whether buyers are distanced too much from reduction efforts by paying in cash rather than in kind, and whether selling emissions rights too cheaply is problematic. He concludes that none of these objections is enough to warrant a ban on emissions trading if it is fairly established, though there might be some justification for restricting trading in certain cases.

Intergenerational justice

Determining the just distribution of our remaining stock of GHGs and how we can trade those rights is an important step towards the necessary reduction of emissions. Yet unless such an agreement mandated very large cuts in GHGs, we are still likely to see global warming to some degree. The larger the mitigation undertaken by us now, the less likely it is that there will be high adaptation costs in the future. Yet assuming, as many do, that future generations will be better off than present generations (just as our generation is better off than past generations), why ought we to make sacrifices now for the benefit of future generations? After all, that would seem to require that the worse off (us) should sacrifice for the benefit of the better off.

In his chapter Kaspar Lippert-Rasmussen argues that luck egalitarianism offers us a plausible account of one way in which the effects of climate change are unjust in the context of intergenerational justice. It is certainly not the only way to view climate injustice and our attendant responsibilities, but it is one that he argues has advantages over some of those found in other distributive principles that have been put forward. He focuses on the standard luck egalitarian claim that it is unjust that some are worse off than others when they have not exercised their responsibility any differently to those who are better off. From a purely luck egalitarian point of view (ignoring non-equality based values), sacrificing now will increase inequality between present and future generations without reflecting differential exercises of responsibility. However, Lippert-Rasmussen offers a reformulation of the luck egalitarian position that claims that individuals who reduce their emissions now to benefit future generations are not subjecting themselves to unjust inequality as they are morally required to act in this way. While he argues that this is not the main reason we have for reducing our GHGs, he claims that a luck egalitarian approach to the distribution of the benefits and burdens incurred by climate change is superior to many of its rivals.

Individual duties and responsibilities

The actions of most states fall well below what is currently required to avert dangerous climate change. In this context where states fail to meet the requirements of justice in relation to climate change, what duties should individuals adopt? Insofar as we assume individuals do have duties, two of the options that they have are to try and make changes to their own lives (for example by living as low emissions lives as they can or going further and trying to reduce their emissions below zero to compensate for the inaction of others), or alternatively they might concentrate their efforts on trying to establish a regime in which states did comply with the ideals of climate justice. Kok-Chor Tan explores the challenges for discharging one's duty in the latter institutional manner. On the institutional account of justice, an individual is required to support and maintain the institutional arrangements of his or her society. In relation to climate change these arrangements will include restrictions on emission-producing behaviour, which an individual will be bound to observe.

Yet the situation faced by most people is that they are not part of an appropriately just regulatory scheme. Most states do not come close to meeting their obligations under a climate treaty which would have a high chance of avoiding dangerous climate change. Within an institutionalist framework, Tan argues that individuals who accept that the scheme in their society is unjust are faced with several options; do nothing, do all one personally can, do what just institutions would require of you if they existed, or help create just arrangements. Tan's argument is that individuals should adopt the latter duty – do what just institutions would require of you if they existed. There are different ways of interpreting and justifying this duty. Tan argues that that individuals have a duty to create just institutions, a duty that is in part explained by the key role that coordinated public action plays in reducing emissions. He notes that a person's direct personal consumption of emissions only captures some of the total amount emitted, with the majority of emissions the result of collective activity such as how public buildings are heated or cooled, what energy efficiency regulations apply or whether there is provision of adequate public transport.

The issue of acting in a context where your state is not meeting its obligations with respect to climate change also raises the general issue of how we evaluate the actions of individuals who participate in groups where the group is involved in wrongdoing. In the case of climate change, one response is to claim that my actions may make no difference to the harms that accompany it, therefore I am not acting morally wrongly by not reducing my emissions. In reply, one might concede that one person's emissions make negligible difference but question whether that is the only sense in which they might be wrong. In his chapter, Garrett Cullity argues that there might be a 'participatory derivation' whereby an individual's act is wrong due to the membership in a group that is acting in the wrong way (a negative derivation), or failing to act when the group is acting as it ought to (a positive derivation). An example of the latter kind is free-riding on the production of an important public good. The actions of the free-rider can be wrong because they are unfair, and this is true even where they harm no one in a direct sense. A negative participatory derivation occurs when an individual joins in an activity that the group ought not to be doing, such as excessive emitting of GHGs.

Yet, on Cullity's account, these two ways in which an individual might act wrongly through being part of a group that is acting wrongly are hard to sustain. Take a case where the deaths of vulnerable people can be prevented through offsetting in a way that is easily affordable for a group. In such a case, failing to contribute to the efforts of your group is placing a burden on others unfairly, or not taking seriously the deaths of others. Yet Cullity suggests several reasons why it is difficult to complain about the actions of an individual in this case –namely, the uncertainty about the value of the action, the cost of performing it to individuals, and the remoteness of the causal contribution. Cullity then argues that the case of groups acting to cause climate change is not like other instances where we act wrongly by participating, for example where I collude with the group to rob a bank, or act negligently to harm someone even if I did not intend to. Cullity claims that this leaves open the possibility that this is a case where the group acts wrongly even though individuals that comprise it do not.

As noted above, one common response to the claim that individuals have duties to make changes to their lifestyles in relation to the climate is 'what difference will reducing *my* emissions make?'. Where the conduct of an agent makes no real difference to the occurrence of the harm, but where he and other agents engage in actions that cause harm collectively, the harm is overdetermined. This is a challenge to those who claim that drivers of SUVs or consumers of meat are harming others through their excessive personal carbon emissions, as their individual actions do not appear to make a difference.

In their chapter, Christian Barry and Gerhard Overland explain some of the reasons for scepticism concerning why these kinds of cases should be of concern. One reason could be that overdetermination cases are troubling because what we lack is a detailed and thorough understanding of the exact causal processes that are occurring. With a sufficiently fine-grained analysis of the causal relations involved in a particular case we can, in fact, determine the actual relation of the person and their action to the harm. However, they argue that this would fail to explain the moral reasons that apply. The argument they put forward is that overdetermination cases are relevant where the outcome is likely to be bad and the probability that by performing an action the agent will be an element of the set of actual antecedent conditions that brings this outcome about. The authors argue that if their characterisation of overdetermination cases is correct, then individuals have a reason to be concerned about their emissions even where they are only a very small part of total emissions, which has implications for how we conceive of compensation claims on the part of those who are harmed.

Harms, agents and belief

Much of the discussion of climate change has focused on various justice related issues regarding how to divide the costs of climate change. While this ought indeed be a high priority, as John Broome reminds us in his chapter, broader ethical issues saturate or should saturate discussion of the problem of climate change. Such issues include ascertaining the nature of climate harms, the agents who should attract moral consideration and our questions concerning how we approach the formation of beliefs about climate change.

For instance, the UNFCC treaty might be said to have three stages: determining what constitutes dangerous climate change, measuring how much GHG we are allowed to emit to avoid this danger and finally dividing these emissions and the associated costs. The issues for the third stage are clearly issues of justice. Yet there are many important ethical issues in stages one and two. For instance, determining what the harm of climate change consists in requires that we calculate what harms we avoid by, for example, reducing emissions drastically and compare them to the costs of doing so or not doing so. Comparing them requires that we put a value on them and that exercise is one that involves ethics.

The focus of John Broome's chapter is the more difficult assessment of the harm of climate change on population, and what to make of the risk of severe climate change causing a collapse in population. Broome notes that certain assumptions about the ethics of population are built into the way in which the consequences of climate change are modelled. While the death and suffering of large numbers of people will be bad in very obvious ways that are not unique to population ethics, another consequence of this kind of severe collapse is that it will *prevent* the existence of many people that might otherwise have lived. A common way of thinking about this issue is to adopt the 'intuition of neutrality', which is to say that adding another person to the world is morally neutral. Broome presents several arguments for why we should abandon the intuition of neutrality, but notes that there is no consensus in population ethics on how to treat such cases. While this presents a difficulty for policy making about climate change that cannot be solved through philosophical discussion, Broome argues that there is nevertheless a role for philosophy in identifying and articulating the relevant questions.

Calculating the type and valuation that we ought to place on different harms to people is a difficult exercise. Yet there is also a question of whether harms should be understood as including harms to the non-human world. In determining whether we should extend moral concern to non-humans, one approach has been to ask whether and what moral status non-humans can be said to have; whether they are rational, experience pain and so on. Extending moral concern in this way obviously makes a large difference to our calculations of the injustice of the harm of climate change, given that many of the impacts will be on other living things. An alternative approach to these issues is to consider instead whether there are justice-generated obligations to non-human nature. This approach is taken by Benjamin Hale in his chapter. He argues that our moral agency comes with the obligation to adhere to certain rules of behaviour that include the obligation to fix problems that we have created. The claim here is that the wrongness of our action in respect to non-human nature is not simply that we do not include other entities on our list of things with moral status, but that our actions are not justified.

Hale argues that disconnecting the question of responsibility from the status of nature and focusing on justification as the defining feature of morality has important results for how we should act in relation to climate change. For one it requires us to justify our actions to a greater degree than we have previously been accustomed. The effects of the actions that are causing climate change, while not intended to harm the planet, are not fully accounted for by the reasons for the actions, which are more limited. Hale's claim is that when we ask whether these actions can be justified, we could not explain our actions in a way that would meet the scrutiny of future generations.

Questions regarding how we come to form our beliefs and treat those with differing beliefs are also relevant to issues of climate justice. One of the tasks that any one person, official or government concerned about climate change may have to undertake is to discuss the importance of action about climate change with a climate sceptic. Scepticism about climate change can exist for many reasons, as Simon Keller points out in his chapter. People can be sceptics because they are lazy about finding the latest scientific statements, have a disdain for science in general or are influenced by well-funded corporations that have an interest in downplaying the significance of climate change. However, rather than dismiss all climate scepticism out of hand, Keller argues that reflecting on how we arrive at our beliefs about climate change and how these intersect with our ideological worldview is important for understanding the sceptic's position. Through a hypothetical example, Keller asks what our response would be if the scientific evidence for climate change suggested mitigating actions that were very harmful to many of the goods that environmentalists hold dear. Instead of actions such as using less fossil fuels, eating less meat or travelling less, what if scientists recommended concreting large parts of our remaining wilderness areas? He suggests that one plausible reaction to this kind of scenario is that environmentalists would find it difficult to accept such claims insofar as they conflicted with their particular worldview – a situation that many climate change sceptics may currently find themselves in. The suggestion is that scepticism can be subjectively rational in the sense that it can be the result of an ordinary person doing the best they can to find out the truth. Keller's point is that what we are doing in part in cases like climate change is choosing between experts, and this can be difficult. The nature of climate science also means that is unlikely that the science will become clear enough to convince a sceptic, and ideological differences are bound to remain. Yet progress in the debate about climate change requires that we address scepticism and Keller provides several suggestions for how this might be achieved.

1 Climate change and state consent

Thomas Christiano

Climate change poses a fundamental challenge to the international community, possibly greater than any it has faced so far. It is a basic challenge because mitigation of the threat appears to require fundamental changes in the economies of most of the world. The economic systems of the various societies in the international community are run on energy that is mostly based on carbon. Mitigation of the threat of climate change seems to require a significant drop in standards of living or a fairly rapid transformation of these economies from carbon-based to some other form of energy source. At the same time the mitigation of the threat of climate change has the strategic structure of a public good. Most societies have incentives, when taking only their own narrow interests into account, to free-ride on the sacrifices of others. And those societies that can make a significant difference to climate change can only do so with the cooperation of others whose cooperation is uncertain. But the challenge arises also because there is a great deal of diversity among societies in their relationships to the causes and effects of climate change as well as the burdens of climate change mitigation policies. Climate change does not appear to affect all societies in the same way and societies contribute to it in different ways. Furthermore, mitigation policies impose different burdens on different societies.

At the same time, the decisions about how to mitigate climate change and how to adapt to it must be made primarily by states acting in concert. The fundamental principle of international environmental treaty law must be state consent. States are the primary players in the making of international law. They remain the centres of power and the most basic mechanisms for the accountability of political power to persons. And states are the entities that are most capable of making the significant transformations that are necessary to achieve the goals of climate change mitigation and adaptation to climate change.

The conventional wisdom is that since the problem of climate change is a global one, the only legitimate method for solving this problem is a process that is fully inclusive and fair to all the parties in the international system. This is the thinking behind the United Nations Framework Convention on Climate Change (UNFCCC) and the Kyoto Protocol. Processes that include only a small number of states are thought of as exclusive and therefore illegitimate. I want to question the conventional wisdom and suggest how more particular and less inclusive methods of treaty construction may in fact be more legitimate than the universal method of the UN framework.

In this chapter I will discuss how state consent can be a genuine basis of legitimacy in the context of decisions about climate change. And I will explore how considerations of legitimacy help us think about some of the different methods of treaty construction in international environmental law particularly as they touch upon the issue of climate change. I will start by explaining the different methods of treaty construction. I will then lay out the objections from legitimacy to what I call the club method. Then I will elaborate and partly defend a democratic and morally bounded conception of state consent as a basis for the legitimacy of international law. I will then address the main problem of the chapter. I will introduce some of the issues of legitimacy that arise when considering the club method and the universal methods of treaty construction. I will challenge the conventional wisdom that the universal method is legitimate while the club method is not by showing under what circumstances the club method may actually be more legitimate than the universal method.

Two basic methods of treaty making

The two basic methods I have in mind are the broad but initially shallow method of treaty construction and the deep but initially narrow method of treaty construction. I will call these respectively the universal method and the club method since this is how they are commonly known. What are these two methods?

The universal method

The universal method is illustrated by the current main approach to treaty construction in approaching climate change. It is exemplified in the UNFCCC and the Kyoto Protocol. The method proceeds by creating a universal or nearly universal agreement among states to a framework convention. The framework convention does not impose any serious demands on societies. It states aspirations and general principles; it defines procedures for making further agreements; it lays out some administrative apparatus and it confirms a scientific process for investigating the vicissitudes of climate change. In this respect the convention is fairly shallow. It doesn't require much from states as it stands but it announces an intention to find collective solutions to the problem of climate change. At the same time the convention is quite broad: it includes nearly every state in the process. In the case of the UNFCCC, this feature is obvious from the fact that it is a UN-based treaty organisation. The framework convention provides for procedures for developing further treaties or agreements that may be more demanding. The UNFCCC provides for the regular meetings of the conference of the parties. These conferences can then develop further agreements that impose requirements on states. The procedure of agreement at least for the basic elements of subsequent treaties is usually that of consensus, although we will need to explore this condition a bit more carefully. The Kyoto Protocol is the main product of this process in the case of the UNFCCC. And the protocol itself imposes deeper commitments, although its commitments were also not very deep, and it too states further areas in which cooperation is to be deepened. But the Kyoto protocol says little about enforcement. The mechanism of enforcement that is in place is very weak. The parties do intend to develop some mechanism of enforcement as time goes along. But they have left this problem to be solved at a later date. Indeed many problems were left for a later date. This is not necessarily a flaw, it is a crucial feature of the method. The idea is to get everybody on board and then slowly deepen the commitments of each party. The aim is to have an inclusive process of decision-making and then to get each to make greater and greater sacrifices over time.

The club method

The club method of treaty making is one which tends to start with deeper commitments among a select group of states.[1] It can then expand outwards to include more and more states by means of accession agreements that new states make in order to join the treaty body. The paradigm case of this method is the General Agreement on Tariffs and Trade (GATT). It began as a relatively small group of states agreeing to significant tariff reductions on particular goods and on the basis of basic principles of non-discrimination among states including the principle of most favoured nation status, which asserts that a state extends the same trade concessions to every member of the GATT that it agrees on with each member and the national treatment principle which requires that states not impose any burden on imported products that it doesn't impose on domestically produced items that are similar (that is, in addition to the tariff bindings agreed upon). The GATT started in 1947 with only 23 member states and has now grown to include more than 153 states by means of accession agreements.[2] The 1947 GATT involved agreements for reducing tariffs among the member states.

---

[1] The term 'club' comes from R. Keohane and J. S. Nye Jr., 'The club model of multilateral cooperation and problems of democratic legitimacy', in R. Keohane (ed.), *Power and Governance in a Partially Globalized World* (London: Routledge Publishers, 2002) pp. 219–44. In this chapter, I am exploring only one aspect of this model as it is laid out by Keohane and Nye, namely the fact that agreements are made within small groups of states rather than requiring a universal process of agreement making.

[2] See B. Hoekman and M. Kosteki, *The Political Economy of the World Trading System: the WTO and Beyond*, 3rd edition (Oxford University Press, 2009), p. 49.

These agreements were already quite significant so they were relatively deep. The agreements made by the member states grew deeper and deeper over time so that they included more tariff reduction and then reduction of barriers to trade and subsidies and many other issues in international trade. From the GATT also came the World Trade Organization (WTO), which included these agreements and then added a more effective enforcement mechanism to the system.

The club method is the traditional method of creation of treaties in international law, at least those created before the creation of the UN. States agree to treaties with particular other states but not with everyone. The international system is at least in part a system of voluntary association in which each state has agreements with some other states and the agreements are regional in character or bilateral or are particularistic in some way. Indeed, most international treaties still have this particularistic character; even most trade treaties tend to be more often particularistic than universal.

We need to distinguish between three different kinds of club methods of treaty creation, which we can already see from looking at the GATT/WTO and comparing it to regional or bilateral trade agreements. Most club agreements are meant to remain club agreements. The African Union, for example, is meant to be a union of African states, it is not intended to be a universal union. It is an exclusive club. But some club agreements are meant to expand into universal agreements. This seems to have been the intention behind the creation of the GATT in 1947. And this difference between club agreements with essentially particularistic concerns and club agreements which have universalistic aspirations is crucial to understanding the difference between the two different methods in the context of climate change. A third type, which we will be discussing in this chapter, is a method of treaty creation that is initially exclusive, in the sense that only some states are invited to join, but that intends gradually to loosen this exclusivity over time to include all or nearly all states.

The problem of climate change appears to require a global solution since the problem seems to involve the existence of a global public bad. The steady accumulation of greenhouse gasses (GHGs) in the atmosphere and the long life of many of these gases, coupled with the fact that the effects of the accumulation of these gases are likely to bring about general changes in the climate system that will affect nearly everyone, suggest that the impact of the behaviours of societies will be quite general. For example sea levels are likely to rise and thus the interests of all societies except land locked ones will be affected usually in some negative way. Combine this with the fact that societies will have a hard time making a significant difference to climate change without the participation of many other countries and could conceivably benefit from many other countries undertaking the burdens of reducing their emission of GHGs and you have strategic structure of interaction, which, taking only the aggregate interests of each of these countries into account, looks a lot like a prisoner's dilemma. A large number of societies, but not all, need to make significant sacrifices to mitigate the threat of global climate change.

One final remark on the club method. I do not mean to imply that this method endorses a single track from a particular club to a universal arrangement. The club method is compatible with there being a number of different clubs developing diverse solutions. So we could see a fragmented group of clubs all trying in their way to make a contribution to climate change mitigation and adaptation.[3]

Legitimacy and the two methods

A number of authors who have discussed these two contrasting methods of international environmental law making have suggested that the universal method is superior from the point of view of legitimacy considerations to the more exclusive club method. For instance, in commenting on the differences between universal and club approaches to climate change policy, David Victor states: 'The legitimacy that comes from giving all nations a voice can be important, but it comes at a cost of much more complicated negotiations that are more prone to gridlock.'[4] And Scott Barrett, with specific reference to climate change agreements, says that 'for reasons of legitimacy, it would be hard to defend an exclusionary treaty ["agreements involving a smaller number of countries rather than a global treaty"]'.[5] Finally, Daniel Bodansky says that 'the global character of the UN climate regime [is] often seen as crucial to the regime's procedural legitimacy'.[6]

Hence, the conventional view seems to be that the universal method is legitimate while the club method is not. Victor and Barrett draw opposing conclusions from their observations. Victor argues that the club method is more effective than the universal method for the kind of problem that mitigation of climate change poses. And he thinks that the greater effectiveness of club-like treaties is worth the sacrifice of the legitimacy that comes with universal treaties. Victor is thinking here of club treaties that are expected to develop into universal or near universal treaties so the sacrifice of legitimacy is not permanent.[7] Barrett argues that 'An effective climate change treaty must be

---

[3] See, for example, R. Keohane and D. Victor, 'The regime complex for climate change', *Perspectives on Politics*, 9/1 (2011), 7–23.
[4] D. Victor, *Global Warming Gridlock: Creating More Effective Strategies for Protecting the Planet* (Cambridge University Press, 2011), p. 50.
[5] S. Barrett, *Environment and Statecraft: the Strategy of Environmental Treaty-making* (Oxford University Press, 2005), p. 394.
[6] D. Bodansky, 'Legitimacy', in D. Bodansky, J. Brunnee and E. Hey (eds.), *The Oxford Handbook of International Environmental Law* (Oxford University Press, 2007), pp. 711–12.
[7] Victor, *Global Warming Gridlock*.

global.'[8] Bodansky, too, though noncommittally, suggests that the global character of the UN climate system, 'may raise transaction costs and make it harder for the regime to achieve results'.[9]

The fundamental problem seems to be that so far the universal method has not produced results and many, including Victor and Barrett, are sceptical that it can produce results.[10] The question this raises for a conception of legitimacy is whether a club method, even if it is significantly more effective, must lack in legitimacy. I want to challenge the conventional view here and I want to do it from within a conception of legitimacy that is animated by democratic principles.

I am interested here in examining and comparing the legitimacy making properties of these two different ways of making treaties on climate change mitigation. I will not here try to assess the relative importance of legitimacy considerations so I will not try to assess the dispute between Barrett and Victor. My concern will be to probe the point of agreement between the two authors. I will ask whether it is true that universal treaties are better from the standpoint of legitimacy than club treaties and I will probe the considerations of legitimacy that seem to be in play.

State consent democratised

We need to make at least four modifications to the traditional doctrine of state consent to make it live up to the idea that it is a conception of legitimacy grounded in the interests of persons in an equal voice in their shared institutional framework. The first three I will mention only for completeness but I will put aside in the rest of this chapter. First, the states that consent must be highly representative of the people in the state. The motivation for state consent is the accountability of states to people. The full realisation of this idea involves the consent of reasonably democratic states. In this way individuals participate in the making of international law through participating in the determination of the positions of the state that is negotiating the law. This way the idea that individuals are bound by international law, directly or indirectly, can be vindicated. The second modification is that the process of negotiation by which treaties are created must be a fair process of negotiation in which the parties are treated as equals in the process. Obviously coercion and fraud are normally ruled out by this standard but so are the kinds of pressure that result from very different levels of economic and military power. Without some way to temper the effects of differences in economic and military power the process

---

[8] Barrett, *Environment and Statecraft*. [9] Bodansky, 'Legitimacy'.
[10] See, for instance, D. Helm, 'Climate change policy: why has so little been achieved?' in D. Helm and C. Hepburn (eds.), *The Economics and Politics of Climate Change* (Oxford University Press, 2011), pp. 9–35; Victor, *Global Warming Gridlock*.

that produces treaties cannot but be seen as greatly favouring the interests of the more powerful parties.

The third modification needed is that we must make room for the role of expertise in the making of international law. What is known as global administrative law in the making of international law has some independence from states, though ultimately they are responsible for it and bound by it.[11] I think that this can be legitimated from within the idea of state consent in much the same way that expertise in democracy can be accorded a legitimate role in the making of law. It does not need to be directly accountable to persons but it must be constrained so that it genuinely and effectively pursues the aims and realises the principles of the principal parties.[12]

The morally mandatory aims of the international community and the limits of state consent

The fourth modification is more important for this chapter and will need a bit more time. The global community is currently facing some fundamental moral challenges, which can be recognised as such on virtually any scheme of morality. The aims of the preservation of international security and the protection of persons against serious and widespread violations of human rights are already recognised in Article One of the *Charter of the United Nations*, which lays out the purposes of the UN. In addition, there are aims of equally great moral importance that must be pursued by the international system. First, it must pursue the avoidance of global environmental catastrophe. Second, it must pursue the alleviation of severe global poverty. And third, it must establish a decent system of international trade. These challenges will require significant cooperation from many of the world's states at least.

The morally mandatory character of the aims and the necessity of general cooperation in the pursuit of the aims imply that there are certain tasks that are morally mandatory for states to participate in. This suggests a set of moral imperatives that are not the usual context for voluntary association. The usual context of voluntary association is that the parties are morally at liberty whether to join or not and even if some associations do pursue morally important aims,

---

[11] See B. Kingsbury, N. Krisch and R. Stewart, 'The emergence of global administrative law', *Law and Contemporary Problems*, 68/3 (2005), 15–62.
[12] I have tried to lay out a theory of how to make democracy compatible with the need for expertise and specialisation in the modern state in T. Christiano, *The Rule of the Many* (Boulder, CO: Westview Press, 1996), ch. 5, and in 'Rational deliberation among citizens and experts', in J. Mansbridge and J. Parkinson (eds.), *Deliberative Systems: Deliberative Democracy at the Large Scale* (Cambridge University Press, 2012). Robert Keohane, Stephen Macedo and Andrew Moravscik have argued that significant independence of international institutions from states can be compatible with democracy in their 'Democracy enhancing multilateralism', *International Organization*, 63/1 (2009), 1–31.

there are enough of them that one may pick and choose among them without moral cost. But the need for large-scale cooperation to pursue morally mandatory aims is what makes the international system a peculiar kind of political system. It relies on consent but cooperation is required to pursue morally mandatory aims.

There is still significant room for the moral liberty that state consent protects in such a society but it must be heavily bound by constraints. The justification for the state consent requirement, and thus some moral liberty to say no, is grounded in the fact that though we are morally required to cooperate in solving these fundamental moral problems, there is a great deal of uncertainty as to how these problems can be solved. Though there is general agreement among scientists that the Earth is warming up due to human activity, there is disagreement as to how much this is happening and how quickly. There is also substantial disagreement about how to mitigate global warming and what a fair and efficient distribution of costs might be. The same uncertainties attend thinking about how to alleviate global poverty, how to create a decent system of international trade and in fact in how to protect persons from widespread human rights abuses.

This kind of uncertainty, together with the centrality of states in making power accountable to persons, provides a reason for supporting a system of state consent with freedom to enter and exit arrangements because it supports a system which allows for a significant amount of experimentation in how to solve the problems. Experimentation within different regional associations as well as within competing global arrangements may be the best way to try to solve many of the problems we are facing. And democratic states are the ideal agents for this kind of experimentation because of their high degrees of accountability and transparency.[13]

But the system of state consent must be heavily bounded given the morally mandatory need for cooperation. In the usual case of treaties, refusal of entry and exit are permissible and require no explanation. In the cases of treaties that attempt to realise a system of cooperation that is necessary to the pursuit of morally mandatory aims, the refusal to enter or exit from it would require an acceptable explanation that lays out the reasons for thinking that the treaty would not contribute to solving the problem and that some alternative might be superior. Exit or withdrawal is permissible but only with an adequate explanation. By 'adequate explanation' I mean an explanation that is not irrational, unscrupulous or morally self-defeating and that displays a good faith effort to solve the problem at hand. The explanation must be in terms of the morally

---

[13] See, for example, Keohane and Victor 'The regime complex for climate change', p. 9, 12 and 15 for a discussion of how uncertainty plays a role in motivating and perhaps justifying fragmentation of regimes attempting to deal with global problems.

mandatory aims or in terms of a crushing or severely unfair cost of cooperation. The explanation need only be adequate not in the sense that it need be the correct explanation, but it must fall within the scope of what reasonable people can disagree on. An irrational explanation goes against the vast majority of scientific opinion. An unscrupulous explanation free-rides on others' contributions to morally mandatory aims or it refuses to shoulder any share in a morally mandatory pursuit. A morally self-defeating explanation is one that insists on a different coordination solution, defeating a coordination solution that in the circumstance advances everyone's aims.

The justification for these constraints on state consent derive from the need for cooperation on morally mandatory aims together with the need for fairness in the pursuit of these aims. Both these considerations are central to determining the legitimacy of a process of creating international law.

Let me situate this claim within the larger framework of international law. The system already recognises that states' agreements are null and void when they consent to something that is in violation of *jus cogens* norms. These pose significant limits to state consent. It seems to me that it is but a short step from these propositions to one that denies states the right to refuse consent to agreements that pursue morally mandatory goals on the basis of the above problematic reasons.

These boundaries of consent pose genuine limits on the sovereignty of states. They authorise other states to come together to pressure unscrupulous states to abandon the strategy of free-riding on the cooperation in pursuit of mandatory aims. Again we see how the international system is a peculiar kind of political system in that states can be authorised to force or pressure other states to cooperate or to provide a reasonable explanation for non-cooperation. These are legitimate and rightful exercises of power on the part of states when in pursuit of morally mandatory aims.[14]

Is the club method of treaty making illegitimate?

My concern then is to look at the claims above from the standpoint of the different grounds of legitimacy. Some have argued that the club method I sketched above suffers from a lack of legitimacy at least relative to the universal method. They have offered only the barest of clues for why they think this. Victor argues that the club method is unfair because it does not give everyone a voice. And perhaps he means to include in the unfairness that not everyone is given an equal voice. Barrett argues that the club method

---

[14] See T. Christiano, 'The legitimacy of international institutions', in A. Marmor (ed.), *The Routledge Companion to the Philosophy of Law* (New York: Routledge, 2012) for further elaboration of these ideas.

is exclusionary. But we need to consider different interpretations of these two concerns.

First I want to consider two obvious objections to the idea that the club method is illegitimate. These authors have not considered the purely instrumentalist conception of legitimacy in their remarks. They both contrast legitimacy with effectiveness in the sense that they both seem to accept the idea that if the non-legitimate process could be carried out, it might be more effective. We will consider the arguments for this view in a moment but here I just want to observe that many have accepted that a process can be legitimate because it is effective. And surely there is something to this. It could very well prove to be the case that a process that is not appallingly objectionable could generate duties on the part of all to comply if it is the case that their compliance with these duties would help promote important moral ends that could not otherwise be promoted. This could well generate content independent duties in many or even the vast majority of participants. If a process tends to produce strong coordination points that might not be possible otherwise, participants might have duties to go along with the directives even if they are not sure that the directives identify coordination points. They simply have duties to go along because of the tendency to produce coordination points.

I will put this point aside because I think there is also something intuitive in the concerns of the authors. The way I understand the concern is that there are different kinds of legitimacy and some are of higher worth than others. My thought is that the legitimacy conferred by consent or by democracy tends to be of a higher worth than the legitimacy conferred by purely instrumental concerns. Here is why. The legitimacy conferred by consent or democracy embodies a much higher grade of moral community than the purely instrumental legitimacy. In the case of consent or democracy, legitimacy involves each person having a right against all the others. Each person does what the directives tell them to do. This means that each person owes the performance to every other person. Moreover both of these accounts also suggest that legitimate decisions are ones of which one can say that the persons are equal and joint authors of the decision. They consent to the decisions or they participate as equals in the creation of the decisions. Finally, consent and democratic accounts of legitimacy ground a peculiarly public conception of legitimacy, such that the parties can see that the legitimate authority is legitimate. In contrast, instrumentally legitimate authorities are such that one does not owe the performance of the duty to them and they are such that their legitimacy is in some sense accidental to the particular powers that have legitimacy. The explanation for their legitimacy is a set of contingent facts that will pass. Furthermore, their legitimacy is not even necessarily something that the participants can see to be in effect. The participants could, at the limit, well regard the instrumentally legitimate power as illegitimate even though it is in fact legitimate.

These points imply an important moral difference between instrumental and consent or democratic conceptions. They point to an intuitively very different kind of community between legitimate power holders and the subjects of those powers. In one community the subjects and power holders are seen primarily as instruments of moral value and many may be inadvertent instruments. In the second instance they are treating each other as equals in accepting their obligations and they are co-authors of their obligations. Furthermore in democratic or consent-based legitimacy, the obligations are generated in a transparent way.

So in response to the first objection, it is certainly true that there can be a kind of legitimacy that is held by a decision-making process on account of its effectiveness. Yet, given the high worth of the democratic and consent-based conceptions of legitimacy, it is worth exploring the extent to which different treaty-making methods differ in their realisation of these standards.

The second objection arises from the context of the traditional doctrine of legal legitimacy for international law, namely the doctrine of state consent. There is a sense in which it is hard to see how the considerations of voice and exclusion are supposed to work against the club method. The basic problem is that where the club method excludes it does not generate any obligations in the excluded participants. And when it does generate obligations it doesn't exclude. And the same holds for the consideration of voice at least on a reasonable interpretation within the doctrine of state consent. But if the club method does not even purport to generate obligations in excluded parties or parties that have no voice and it only purports to generate obligations in non-excluded participants, then it is hard to see how it can suffer from defects of legitimacy or illegitimacy.

Let me explain this a bit more. Let us articulate three different stages of agreement for the club method. The first is the stage of no agreement. The second is the stage of limited agreement among club members. The third is the stage of complete agreement among all potential participants. This sequence is envisioned by the authors. In particular, David Victor says that climate change agreements must start with agreements among a select club of societies that are most enthusiastic or that contribute most to climate change and then subsequent societies may join by means of 'climate accession deals'.[15] Ultimately he envisions virtually every society in the world joining either through the initial club or through subsequent accession agreements. The process he envisions is much like the process that led to the creation of and then expansion of the GATT and then the WTO. Except here only a small number of states are permitted to join and at a certain point later in time, almost everybody is a member.

---

[15] See Victor, *Global Warming Gridlock*, p. 24.

Let us call the second and third stages of agreement the club and universal stages. Initially one might think that there is no danger of illegitimacy at any one of these stages. At the first stage there is no legitimacy but no defect either. No one is obligated to do anything as a result of international law. At the second stage there is no defect either since only those who have joined the club are obligated. Those who have not signed on to the agreements are not obligated. At the third stage nearly everyone is obligated but that is because they have all signed on. So where is the problem?

Legitimacy and illegitimacy in international environmental law

In response to this objection, I believe that the inference from the claim that legitimate authority imposes obligations to the claim that only when someone purports to impose obligations does the issue of legitimacy arise, is mistaken. There is another important basis of illegitimacy. There are contexts in which there is a demand for common action and there is disagreement and conflict of interests as to how best to structure the common action. When an agent acts unilaterally in violation of a demand for legitimate decision making, we can say that the agent acts illegitimately even when no obligations are purportedly imposed. For example, if an agent tries to secure a common action merely through coercion, then the agent acts in violation of the demand for legitimate decision making, and thus acts illegitimately even if there is no purport to impose obligations. And, if the agent imposes certain kinds of costs on other parties that it may only impose under the condition that the parties consent or have a voice, then that party acts illegitimately. And the costs imposed need not include any purported obligation. We have illegitimacy through costs without voice or consent. Let me explain this further.

The issue of legitimacy can arise just when one person imposes serious costs on another. The question 'by what right do you do this?' can arise when one party purposefully makes use of another's property or imposes a serious cost on it like in cases of eminent domain, quartering soldiers, taxation or pollution. And the answer that it is done for a good public purpose is not sufficient without a claim of legitimacy. To be sure, individuals can make use of other persons' property in cases of dire necessity or in cases of easement where the damage is very superficial. But we are talking here of making use of property or imposing significant damages merely for the purpose of advancing the public good.[16]

Another case like this occurs in the international community. For instance, when one country imposes significant costs on another through pollution, it is

---

[16] See H. Grotius, *On the Law of War and Peace*, student edn, S. Neff (ed.), (Cambridge University Press, 2012), p. 226 for this conception of the right of eminent domain.

expected that the affected country is to be called in to work out some arrangement for mitigating or compensating for the costs. There is a generally recognised 'duty to consult and to negotiate in good faith with neighbouring states about possible trans-boundary pollution'.[17] When acid raid was discovered to be coming from the US into Canada and from some European states to others, these countries entered into agreements to solve the problem. I think we can make sense of this duty as a way of assuring the legitimacy of decisions regarding how to deal with the costs that societies impose on others in a way that takes into account all of the interests that are at stake.

It is sometimes thought that states merely have duties to prevent such pollution, but it does not seem to me to make sense to say this in all cases. If the pollution is the result of a highly beneficial industrial process, it may be irrational both from the standpoint of the state involved and from the standpoint of the international community to eliminate the process. What may be needed is some mix of prevention and redistribution of costs. What is called for is common action in the sense that the polluting state and the neighbouring states ought to negotiate a fair and sensible solution to the problem. The call for common action is accompanied by a call for legitimacy in the decision making behind the common action. Hence, the consent and democratic norms will play a role here in determining what a legitimate way of making the decision should be.

The imposition of pollution by one country on another without consultation or negotiation is a kind of illegitimate exercise of power of one over the other. The imposition of pollution itself certainly cries out for some kind of legitimate way of making decisions about how the costs and benefits of this activity are to be arranged. That this is so is confirmed by the existence of the general principle of international law we cited above. It gives rise to the demand for legitimate decision making. But we can also argue for this more substantively. When one country imposes pollution on another it sets back the interests of the other country for the sake of pursuing its own interests. It, in effect, realises a certain distribution of burdens and benefits among societies. And it does so unilaterally. This distribution is likely to be quite controversial at best. Each society has a serious interest in having a voice in determining the best way to deal with the pollution. For one country to pollute another country is for it to decide how the distribution of benefits and burdens are to be distributed between them without their having a voice. And this is what legitimate decision making is supposed to deal with. If one country pollutes unilaterally, it simply usurps the legitimate decision-making power that should be held by the two countries together. So I think we can correctly say that an institution or

---

[17] See D. Bodansky, *The Art and Craft of International Environmental Law* (Cambridge, MA: Harvard University Press, 2010) p. 95.

decision-making process suffers from a defect in legitimacy if it imposes significant costs on others without their consent or voice. What I will try to show eventually is that the club method of making agreements among states suffers from a serious defect of legitimacy if the agreements impose costs on non-participating states.

But how do we avoid the difficulty described by Nozick of implying that the rejected lover has a right to a say in the beloved's new love life?[18] I am not sure we can answer this question entirely. But it may be that the best way to think of this is that there are certain protected interests that persons have, which are such that they do not owe anything to those who are harmed directly by the satisfaction of those interests. The interests in love, association, conscience and expression of ideas are among these. These are interests such that the realisation of these interests in action are protected from many kinds of interference from others. The interests of individuals that I have in mind here are interests in intimate relations with others, friendship and interests in relationships that touch on one's most fundamental commitments about what a good life consists in. These interests are protected by stringent rights of conscience, association, expression and privacy, which are such that no democratic legislature may abridge the core of them without extraordinary reason.

Beyond these protected interests, individuals have interests in having a say concerning the social and economic environment they live in. This is what grounds a democratic right to a say in defining property rights and exchange as well as taxation and education. Certainly there are many damages that occur in social life that society chooses to ignore on the grounds that they are part of a scheme of cooperation from which everyone benefits. Some losses that might occur from market competition are among these. So, as Mill says, though there is a demand for legitimate authority in the case of these losses, it may be that the democratic authority deems them acceptable.[19]

Damages to persons that concern unprotected interests can be legitimised through a democratic process in domestic society. The imposition of damage by one person on another can be legitimised by the fact that, as citizens, they have had a say in the democratic process in whether those damages are acceptable parts of social cooperation. For example, individuals do have some say concerning the damages that arise out of market competition. This say is through the democratic process. And some damages are mitigated such as loss of employment through unemployment insurance. Others are normally not mitigated such as loss of market share as a result of one's competitor making a more desirable product. Each has a say in this process and the outcome is that some losses are blocked, some are mitigated and others are left alone. In these cases

---

[18] R. Nozick, *Anarchy, State and Utopia* (New York: Basic Books, 1974), p. 189.
[19] See J. S Mill, *On Liberty* (Buffalo, NY: Prometheus Books, 1986), p. 107, ch. 5.

individuals do not have a say directly over the damages that might occur and they do not have to consent to the loss of market share or even to unemployment. But they do have a say, through the democratic process, over the rules that regulate the market, which in turn regulates the resulting costs of market interaction. There is a kind of division of labour concerning how to deal with the costs of ordinary interaction.

What is interesting about international environmental law is that this two-level structure of interaction does not occur, at least usually. When one state imposes damages on another, there is not a larger political framework that legitimises it. There is no separate political institution to which states can appeal to determine whether these damages are appropriate or not.[20] In the context of international community, it seems to me that we can make sense of the duty to negotiate in good faith mentioned above as a kind of call for common action to deal with a trans-boundary pollution problem.

This is, I think, a clear and rightful limitation on the sovereignty of a country. A country may not develop its industry or its physical environment in such a way that benefits itself and pollutes others without consultation and negotiation with those others. This does not require that a country desist with the activity, but it does require good faith negotiation to work out a mutually beneficial and fair arrangement. From this we can see that though there is a limitation on sovereignty here, it is not a simple limitation. A country may exercise its sovereignty by negotiating a good deal with the other affected countries, though it must engage in fair negotiation with them.

The foundation of this requirement, I assert, are the interests of all the people in all the affected countries. A political system of international law that did not assert as a general principle that societies must negotiate mutually agreeable terms with those countries on which they impose damages would be a system that failed to take seriously the interests of the people in the victim countries. It would fail in much the same way that a more centralised political system would fail if it were not to give a right to vote to some significant part of its population.

External effects without voice

There are two distinct problems here that need some discussion. One problem is that the club treaty arrangement may intentionally impose the costs on non-members at what I have called the club stage. The second problem is that the club method involves unfairness in the process of making the ultimate agreement at the universal stage.

---

[20] This is a slight overstatement since there is a well-known case (*Trail Smelter*, Arbitral Tribunal, 1939) in which an international arbitral tribunal ruled that Canada had a duty to prevent transboundary pollution into the US. Even here, both states had to agree to arbitration.

Many treaties impose external effects on non-members so as to ensure that the members receive sufficient benefits to keep the members complying and to encourage non-members to join.[21] A prominent example of this kind of external effect is a trade restriction on non-members. The Montreal Protocol and a number of other treaties permit members to impose restrictions on trade with non-members. In the case of the Montreal Protocol, it permits restrictions on trade of goods that are made with chlorofluorocarbons (CFCs) with countries that are not participating members of the treaties. The purpose of this is to create a benefit for the participating members and to stop leakage of business to countries that do not restrict the production of CFCs.[22] But this does impose a cost on non-participating countries.

In the case of the Montreal Protocol, there may be no real problem of legitimacy. There are two reasons why these trade restrictions are not a fundamental problem. First, all states are invited to participate in the treaty, so there is a sense in which a state has a real opportunity to avoid the sanctions. Second, the imposition of the trade restrictions is designed in part to get more states to participate and abandon the strategy of free-riding. I have argued above that this is a permissible strategy for states to engage in when it comes to the pursuit of morally mandatory aims. States may impose sanctions on those who free-ride on the pursuit of morally mandatory aims. Though the system of international law is a system properly based on the consent of states, that consent may not be withheld for the purpose of free-riding or some other strategy obviously inconsistent with the duty of a state to cooperate in the pursuit of morally mandatory aims. States may withhold their consent from a treaty designed to pursue a morally mandatory aim only if its withholding is based on reasonable disagreement with the appropriateness of the design of the treaty in pursuing the aim. Mere free-riding is not a permissible basis for withholding consent. And I have argued that other states are permitted to impose penalties on states who do merely free-ride. The trade restrictions imposed by the Montreal Protocol on non-participating states can be justified in this way.

We can see that this claim involves a modification of the proposition I advanced above, which stated that when one state imposes a cost (in terms of non-protected interests) on another state without that other state's having a voice in the matter, the imposition is an illegitimate exercise of power. The modification is that the exercise of power is illegitimate unless the imposed upon state is unreasonably refusing consent to cooperation in pursuit of a morally mandatory aim and this is designed to get the imposed upon state to consent.

---

[21] See Barrett, *Environment and Statecraft*, p. 320.

[22] See Barrett, *Environment and Statecraft*, p. 313. Another example of this is *The North Pacific Fur Seal Treaty* (1911) which forbids the importation of sealskins taken from the North Pacific by non-participating countries into the participating countries (Article III).

The situation in the case of the club model is a bit more complicated, however, than the case of the Montreal Protocol. The club model endorses an initial exclusivity in the creation of the treaty. Other states are not welcome to join the club in the initial years of the treaty. Or at least this is how advocates think of it. And advocates usually have in mind that some trade restrictions are to be imposed on trade with non-participating members in order to avoid undermining the effect of the treaty. Otherwise, greenhouse gas producing industry could simply avoid a tax or a cap on greenhouse gas emissions (imposed by participating states) by moving to another country that does not have the tax or the cap because it does not belong to the treaty. Trade restrictions with such countries give incentives to these firms to stay within the society that is capping emissions. And thus trade restrictions may be essential to the effectiveness of the treaty. But if trade is restricted with a non-participating society, then that society is likely to be harmed as a consequence of not being a member of the treaty. This, coupled with the fact that society is not permitted entry into the treaty, seems to be a defect in the legitimacy of the treaty.

A system that imposes restrictions on states so that they or their members do not engage in such free-riding on the pursuit of a morally mandatory aim seems to me not to suffer from any defect of legitimacy. But this is in order to get these states to consent. If however, the treaty is exclusive, there is no effort to get the state to consent. All that occurs is that the state has a cost imposed on it by an arrangement to which it does not have the opportunity to consent. So this case is a kind of intermediate case.

In the extreme case it is permissible for other states to pressure recalcitrant states into participating and cooperating in the treaty-making process when it is in pursuit of a mandatory aim and there are no reasonable objections. This gives the recalcitrant state some kind of say although it is limited. The other extreme case is that it is wrong to impose costs on states that do not consent to them and that are not for the sake of mandatory aims. In the intermediate case we are looking at we are thinking about whether it is permissible to impose restrictions on states for the sake of the pursuit of a mandatory aim but in which we are not asking for the consent and participation of the state in question or its members (indeed we are explicitly excluding it). Despite all these facts we are nevertheless imposing a cost on a state that it is not invited to negotiate about. It is this imposition of cost without prior invitation to participate that makes for a defective legitimacy.

I want to remark on the nature of the damage I am describing here. I argued that when one country pollutes another country without consultation and negotiation with that country, that country thereby acts illegitimately. I am now saying the same thing about one country imposing trade restrictions on another country without that country's consent. To impose trade restrictions on another country without negotiating with that country has a problematic legitimacy. Why? The quick argument for this simply takes the background conditions of the GATT/WTO for granted and the general norms of very low trade barriers between countries. These norms set the baseline for damages. The trade restrictions are generally a kind of breach of the norms agreed upon by states. And to the extent that they make the restricted country worse off, they are an imposition of burdens on them. And these restrictions are controversial.[23] Hence, unilateral restrictions on trade that are not sanctioned by a regime to which both countries agree, and that are not accompanied by an invitation to participate, are illegitimate and problematic.[24]

Path dependence and unfairness in the club method

There is a second possible source of problems for the legitimacy of club-like arrangements that are meant to turn into universal arrangements. These derive from a standard by-product of making decisions in the way the club method envisions. What I have in mind here is that the club method envisions a limited and exclusive set of societies making the initial agreements. After some suitable time, other countries may be invited to join the arrangement by means of accession agreements. The obvious consequence of this arrangement in many cases will be that the earlier societies will normally have a much greater say in the construction of the overall agreement even at the end of the process than the societies that join later. The idea behind this is that the process by which these agreements take place will exhibit a kind of path dependence. The content of the later agreements will be in significant part determined by the earlier agreements. Those who participated in the early stages play a more significant role in the formation of the agreements than those who join later. Now this effect has a problematic legitimacy because the treaty-making process is in some sense designed to have this effect. The process is designed to give incentives to the early participants to join early. And others are excluded from this process. This is of course very different from a process in which all are invited to join at the beginning and only some do, and in which that small initial number plays a large role in determining the content of subsequent arrangements. In this latter case, though, there is path dependence, it is open to all.

In effect the process is set up to give the early joiners more say. And since the others are temporarily excluded, the inequality in say is a part of the design.

---

[23] See Bodansky, *The Art and Craft of International Environmental Law*, p. 249; Barrett, *Environment and Statecraft*, p. 322; and Keohane and Victor 'The regime complex for climate change', p.18.

[24] As when the Dispute Settlement mechanism authorises a state to impose restriction on another state that has violated its agreements.

This may be a problem of legitimacy as well. To the extent that we invoke a democratic conception of legitimacy that values equality of say in the process of collective decision making, and some are given a lesser say by design, there is reason for complaint.

But there are some potential qualifications to this worry. The first one is that the democratic principle does not require equality under all circumstances. And these circumstances may be relevant here. What I have in mind is a principle that requires a say in some collective enterprise that is proportionate to one's stake in the enterprise. Other things being equal, to the extent that one's stake in some cooperative arrangement is greater than others' stakes, one may permissibly have a greater say in the determination of the enterprise.[25] The principle of an equal say is a normal principle in the case of collective decision making in the state because one normally thinks that people have overall roughly equal stakes in the operation of their state. But when one is dealing with particular issues or particular complexes of issues that are separated from other issues, one might think that societies tend not always to have equal stakes in the arrangements. This is true of individuals in political societies as well. They do not have equal stakes in each issue or even each small group of issues; they have equal stakes in the whole complex of issues that face the state. And they can trade off issues in which they have less stake with others who have more stake in them, for the sake of issues in which they have more stake.

Climate change is likely to involve a set of issues in which different societies have different stakes. And it is not unlikely that some of these societies have less of a stake in climate change and climate change negotiations than others. Furthermore, there may be some reason to think that the states that must be included in the early negotiations are the ones that are most likely to have the greatest stakes in climate change negotiations. Presumably one of the main reasons why it is important for these states to participate is that they each have a significant enough interest in mitigation that they are willing to participate. Furthermore, they are states with sufficient capacity to do something about climate change, and that make a sufficient contribution to climate change, that their actions will make a significant difference to climate change.

Assessment of the legitimacy of the club method

It does seem to me that there is some defect in the legitimacy of these arrangements to the extent that the excluded state and its members are expected to cooperate on terms that they have played no role in elaborating. But the defender of the club arrangement might still be able to salvage the legitimacy

---

[25] See H. Brighouse and M. Fleurbaey, 'Democracy and proportionality', *Journal of Political Philosophy*, 18/2 (2010), 137–55 for a discussion of this broader principle.

of the arrangement in the following way. The defender of the club arrangement might argue that there are really two basic alternatives: a universal system which accomplishes little or nothing and a club arrangement that accomplishes something significant but that does so by giving more say to some states and their members than others. The exclusion of states is temporary so they do eventually have some kind of restricted say in the climate accession agreements by which they join the treaty. The defender of the club arrangement will concede that there is some illegitimacy in this way of solving the climate change problem. But, the defender can also say that there are even greater legitimacy problems with the alternative. In effect the alternative deprives all societies of a say over the problem of climate change (under the assumption that it accomplishes little or nothing). Hence, the comparison might be between a decision-making process in which societies and their members have some power over how to solve the climate change problem though a highly unequal power, and one in which states and their members have little or no power to solve the problem.

We might then say that what we are faced with in the universal system is a levelling down problem. The consequence of including all states (so the defender of the club method asserts) is that no state has much if any power to deal with the problem. Even the states that have significantly less say in the club method may have more, or at least not less, say than in the universal method.

Now here is the main argument. Legitimacy comes in degrees. It must come in degrees because the grounds of legitimacy come in degrees and the consequences of legitimacy come in degrees. Legitimacy results from a process of voluntary and fair agreement making and issues in reasons for action. But voluntariness comes in degrees in the sense that parties can feel more or less pressured to enter agreements, and they can have varying amounts of understanding of the nature and consequences of the agreements they enter into. In addition, I want to say that legitimacy depends on the fairness of the process in part. But processes of decision making can be more or less fair. Furthermore, legitimacy issues in reasons for action. And reasons for action can come in different weights as well. If the grounds and consequences of legitimacy come in degrees, it is hard to see how legitimacy does not come in degrees. To be sure, there may be a minimum threshold of voluntariness and fairness beneath which there is no legitimacy, but above that threshold there will be differences of degree.

Now fairness in the distribution of power to solve the problem decreases when everyone has less capacity to exert some influence over the outcome. A decision-making process is more fair than a second one if everyone has more capacity with respect to solving the collective problem than in the second one. This is based on the desire to avoid levelling down in a conception of fairness. Fairness, after all, matters because the things that are distributed fairly are things that matter to all the participants.[26] The point of the decision-making process is to make some progress in mitigating and adapting to climate change. The decision-making process gives parties resources with which to influence that outcome. And fairness in the decision-making process on these issues requires a fair distribution of resources with which the parties can influence this change. So, a concern for fairness in the distribution of a resource or capacity implies a concern that the parties have more of these resources or capacities rather than less. Hence it makes sense to think that a system that grants greater power to all participants is more fair than a system that gives less to all.[27]

The universal method grants little or no effective power to anyone, while the club method grants at least some effective power to everyone though it grants more to some than to others. Hence, though the club method, with its unconsented to restrictions on others and its designed inequalities of say, is not fully fair, it is not as unfair as the universal system, which grants no effective power.

Now, the legitimacy of a decision-making process, I want to say, is a function in part of the fairness of the process. So a decision-making process that is more fair is, other things pertaining to legitimacy being equal, more legitimate. So, the legitimacy of the club method is greater than that of the universal method even though it is defective.

Now I want to say here that this result honours the fundamental intuition that there is something problematic with respect to legitimacy in the club method and that is connected with its unfairness and exclusivity. But it also suggests that a more equal but ineffective process may be even less legitimate. And so while honouring the basic intuition it also rejects the conventional view that if we are concerned with legitimacy, we ought to pursue the universal method.

To be sure, all of this depends on the hypothesis, accepted by proponents and opponents of the club method, that the universal method is highly likely to be fruitless.[28] I am not really in a position to provide an evaluation of this claim. It is not implausible but I cannot demonstrate it. I content myself in this chapter with arguing for the claim that if the club method actually succeeds in making some significant difference to the problem of climate change and the universal

---

[26] See T. Christiano and W. Braynen, 'Inequality, injustice and leveling down', *Ratio*, 21 (2008), 392–420 for a fuller argument for the thesis that the correct understanding of a principle of equality prohibits levelling down.

[27] An analogy of this reasoning applies to the choice between representative democracy and direct democracy. Some may think that direct democracy grants a more equal distribution of ability to influence a decision than representative democracy. But representative democracy, with its division of decision-making labour, is far more effective and this is often for everyone involved. Hence I would want to say that representative democracy is more fair than direct democracy.

[28] I have been working with a pure form of this hypothesis in which it is certain or nearly certain that these facts obtain. Once we introduce significant probabilities that these facts do not obtain, the conclusion becomes uncertain.

method continues to make little or no difference, then there are reasons of legitimacy, within the democratic and bounded state consent conception of legitimacy, to favour the club method, despite its drawbacks.

Conclusion

I have argued that if the proposition that the club method gets things done and the universal method does not is true, then we may reject the conventional wisdom that the universal method is legitimate while the club method is not. I have argued that under these conditions the club method is more legitimate than the universal method even on the democratic and bounded state consent conception of legitimacy. I have attempted to ground this thesis in an argument about the nature of legitimacy in the international system and some general propositions about fairness and its relation to legitimacy. I have not attempted to evaluate the empirical proposition on which this conditional depends though I think it is at least plausible (certainly in the light of the failures of the universal process so far). Furthermore, I have not said anything about the legitimacy of a process in which the antecedent does not hold. My guess is that an unequal distribution may be more fair than an equal one in which the parties are all or nearly all worse off even if there is some small number of parties that are a bit better off under the equal one. But this requires more space than I have here to elaborate.[29]

[29] I want to thank Andrew Williams, Simon Caney, Victor Tadros, Stefan Sciaraffa, Jeremy Moss and Violetta Ignieska for helpful comments on a previous draft of this chapter.

2 Geoengineering in a climate of uncertainty

Megan Blomfield

Against the background of continuing inadequacy in global efforts to address climate change and apparent social and political inertia, ever greater interest is being generated in the idea that geoengineering may offer some solution to this problem.[1] The definition of this concept is itself a matter of dispute, but roughly speaking, geoengineering can be understood as the intentional, large-scale modification of the Earth system. Following the 2011 Expert Meeting on Geoengineering of the Intergovernmental Panel on Climate Change (IPCC), this subject was discussed by all three working groups of the IPCC for the first time in its latest report. Given that geoengineering is an ethically and politically contentious topic, some might worry that this is a sign of its entering the mainstream.

A diverse range of techniques are often placed under the heading of geoengineering. In this chapter, I start by offering a particular definition of the concept and looking at how such techniques are situated in terms of the different responses that are available for addressing climate change. Though many of the ethical and political issues raised by geoengineering remain unsettled – and international institutions for the governance of geoengineering activities are yet to be established – attempts are frequently made to justify further scientific research into the possible impacts of deployment. Lab-based research in particular – investigating geoengineering through the use of models, computer simulations and contained experiments – is perceived by many scientists to be ethically unproblematic. In fact, some advocate a 'research first' approach, which would see extensive lab-based studies conducted *before* international institutions for the governance of geoengineering activities are established.[2] One might think that by improving our understanding of these

---

[1] My thanks are due to Clare Heyward for her helpful discussion of this topic, and my paper, with me; and to Tamsin Edwards for patiently answering my confused questions about the science of climate change and geoengineering. In thinking about climate science and uncertainty, I have benefited from discussions with Ken Binmore, Jonty Rougier and Erica Thompson. It should go without saying that the argument presented in this chapter reflects my position alone, and that any mistakes are my own.

[2] On the 'research first' proposal, see S. M. Gardiner, *A Perfect Moral Storm: the Ethical Tragedy of Climate Change* (Oxford University Press, 2011), 349ff.

techniques and the issues that they raise, such research will enable us to determine the appropriate form of geoengineering governance.

I am not going to take a position, here, on whether or not geoengineering could ever be morally justifiable. My goal in this chapter is more modest – but also has broader implications. I aim to show that *even if* some form of geoengineering might be ethically acceptable in certain specific circumstances, lab-based research into such techniques could nevertheless have morally problematic consequences. I support this claim by explaining that our current state of uncertainty regarding how the impacts of geoengineering interventions could be geographically distributed may help to promote international agreement on fair rules for the governance of geoengineering. In these circumstances of scientific uncertainty, international actors also face uncertainty regarding who the winners and losers could be with respect to potential rules of geoengineering governance, thereby obstructing the pursuit of self-interest in the selection of such rules. Instead of a research first approach, then, we have reason to take a *governance first* approach – ensuring that fair international institutions to regulate geoengineering activities are established *before* further research is conducted into how the costs and benefits of such interventions could be distributed.

Climate change and geoengineering

In order to understand why geoengineering is sometimes proposed as a solution to climate change, it is important to know how climate change takes place. Climate change results from the enhanced greenhouse effect, created by increased atmospheric concentrations of greenhouse gasses (GHGs) such as carbon dioxide (CO_2) and methane (CH_4). Of the solar radiation reaching the Earth system, roughly 30 per cent is reflected back into space and about 20 per cent is absorbed by the atmosphere. The rest is absorbed by the Earth's surface and emitted outwards again in the form of long-wave radiation, and the greenhouse effect takes place when GHGs and clouds act like a blanket to prevent some of this long-wave radiation from escaping, trapping heat in the lower atmosphere.[3] GHGs are added to the atmosphere via a number of natural processes (such as respiration and decay). They are removed from the atmosphere by mechanisms termed *sinks* and may be stored (sequestered) in reservoirs. For example CO_2, the most significant anthropogenic GHG, is sequestered by the ocean, vegetation and soils.[4]

---

[3] See IPCC, *Climate Change 2013: the Physical Science Basis. Contribution of Working Group I to the Firth Assessment Report of the Intergovernmental Panel on Climate Change.* T. Stocker *et al.* (eds.), (Cambridge, UK and New York: Cambridge University Press, 2013), pp. 126–7.
[4] *Ibid.*, p. 544–5.

Due to human activities, GHGs are now being added to the atmosphere much faster than they are removed by natural processes. Two major contributing factors to this accumulation are the burning of fossil fuels (which releases GHGs into the atmosphere) and deforestation (which reduces the amount of atmospheric CO_2 that is sequestered in vegetation – vegetation sometimes referred to as *biomass*). By reducing the amount of long-wave radiation escaping into space, the enhanced greenhouse effect alters the Earth's energy budget, with significant consequences for the climate system. That the climate system has warmed is 'unequivocal', with many of the changes observed since the 1950s 'unprecedented over decades to millennia'.[5] This warming can be expected to have various knock-on effects for the climate, including: sea level rise, changes in atmospheric and oceanic circulation patterns, increased incidence and intensity of precipitation, droughts, heat waves and other extreme weather events. Increased atmospheric concentrations of CO_2 also lead to ocean acidification.[6]

The two major options for dealing with the problem of climate change are mitigation and adaptation. Mitigation is commonly taken to refer to 'a human intervention to reduce the sources or enhance the sinks of greenhouse gases'.[7] Such interventions include burning fewer fossil fuels; developing carbon capture and storage (CCS) capability; reducing the burning of biomass that accompanies deforestation; preserving existing forests and planting new ones. Adaptation, on the other hand, refers to an 'adjustment in natural or human systems in response to actual or expected climatic stimuli or their effects, which moderates harm or exploits beneficial opportunities'.[8] Examples of adaptations include coastal protection infrastructure, crop and livelihood diversification, insurance, water storage, disaster responses and migration.[9]

In three out of the four hypothetical scenarios for future atmospheric GHG concentrations considered by the IPCC in its latest report, global surface temperature by the end of the twenty-first century was found *more likely than not* to exceed 2°C (the internationally defined limit beyond which anthropogenic interference with the climate system is said to become 'dangerous'), and in two of those scenarios it was judged *likely* to exceed this target.[10] Only in the stringent mitigation scenario was warming deemed *unlikely* to exceed 2°C, and

---

[5] *Ibid.*, p. 4. [6] *Ibid.*, pp. 79–113. [7] *Ibid.*, p. 1458.
[8] IPCC, *Climate Change 2007: Impacts, Adaptation and Vulnerability. Contribution of Working Group II to the Fourth Assessment Report of the Intergovernmental Panel on Climate Change*, M. L. Parry et al. (eds.), (Cambridge, UK and New York: Cambridge University Press, 2007), p. 869.
[9] *Ibid.*, 721–2.
[10] See www.unfccc.int; IPCC, *Climate Change 2013*, p. 102. The following expressions, used by the IPCC to state its evaluation of the likelihoods of certain outcomes, should be read as follows: *likely* (66 to 100 per cent); *more likely than not* (>50–100 per cent); and *unlikely* (0–33 per cent). Italics are used to indicate that these terms should be understood in a technical sense (*Ibid.*, p. 36).

this scenario would require that we eventually achieve *negative* global emissions through the use of large-scale bio-energy with carbon capture and storage (BECCS) – a method that some would class as geoengineering.[11] These projections thus help to explain both why increasing interest is being generated in the idea of geoengineering, and why some have begun to worry about the normalisation of such techniques.[12]

Defining geoengineering

How to define geoengineering is a matter of dispute, but the term itself offers a guide to its meaning: the noun 'engineering' referring to 'the branch of science and technology concerned with the development and modification of engines (in various senses), machines, structures, or other complicated systems and processes using specialised knowledge or skills'; and the prefix 'geo' indicating that such development and modification relates, in particular, to the complex system that is the Earth.[13] Geoengineering is often taken to refer only to modifications of the Earth system – or the climate system – that are *designed to address the problem of climate change*.[14] I instead allow geoengineering to be understood more broadly, taking it to refer to *any* intentional, large-scale (regional or global) modification of the Earth system. Geoengineering is almost always discussed as a means by which to counteract climate change, but such interventions could conceivably be designed some other reason – for example, as a method of warfare.

In the context of climate change, geoengineering proposals are often divided into two sub-categories: solar radiation management (SRM) and carbon dioxide removal (CDR).[15] CDR techniques aim to reduce atmospheric concentrations of CO_2 by increasing the uptake of natural sinks and reservoirs (terrestrial and oceanic) or engineering new systems that will remove CO_2 from the atmosphere. Examples include: enhancing terrestrial sinks through afforestation and reforestation; enhancing marine sinks through ocean fertilisation (to

---

[11] *Ibid.*, p. 526.
[12] See, for example, ETC, 'IPCC and geoengineering: the bitter pill is also a poison pill', ETC News Release, 16 April 2014. Retrieved from: www.etcgroup.org/content/ipcc-and-geoengineering-bitter-pill-also-poison-pill (last accessed 20 May 2014). I do not intend to imply that a desire to prevent dangerous climate impacts is the only – or the predominant – reason why geoengineering proposals are receiving increasing levels of attention. Some parties will be attracted to geoengineering because it provides a diversion that might help to delay meaningful action on climate change, or distract from the fact that such action is not taking place; because it offers a way to make money through patented technologies; because they think it might prolong the ability to use or profit from fossil fuels; or simply because they find it an interesting idea.
[13] See the Oxford English Dictionary (www.oed.com).
[14] See, for example, J. Shepherd *et al.*, *Geoengineering the Climate: Science, Governance and Uncertainty* (London: Royal Society, 2009), p. 77; IPCC *Climate Change 2013*, p. 1454.
[15] See Shepherd *et al.*, *Geoengineering the Climate*, p. 1.

encourage phytoplankton growth and carbon uptake); biomass sequestration (where the carbon drawn down by vegetation is stored, for example by burying or conversion into charcoal); BECCS (where biomass is instead used as a fuel and the resulting CO_2 is captured and stored); and industrial processes that capture CO_2 from the air.[16]

SRM, on the other hand, does not reduce atmospheric concentrations of GHGs. SRM techniques are instead designed to reflect more incoming solar radiation back outward into space, thereby preventing global or local temperatures from increasing as much as they would otherwise for any given increase in atmospheric concentration of GHGs. Proposals include preventing as much solar radiation from reaching the Earth by positioning sunshields in space; or increasing the Earth's albedo (that is, the fraction of solar radiation that it reflects) by painting roofs and other surfaces white, increasing crop reflectivity, covering deserts with reflective surfaces, whitening clouds and injecting particles into the stratosphere to scatter more sunlight back into space.[17]

It is important, however, to note that the set of geoengineering proposals for responding to climate change cannot simply be defined as the collection of all CDR and SRM methods. First, CDR and SRM techniques are not always of sufficient magnitude to plausibly count as geoengineering – small-scale afforestation and surface-whitening being two examples. Second, not all potential geoengineering responses to climate change fall into one of these categories. Techniques may be sought to remove other GHGs (such as CH_4) from the atmosphere, suggesting that we might want to consider a broader category of greenhouse gas removal (GGR).[18] Another idea that doesn't fall into either category is 'cirrus thinning', which seeks reduce the greenhouse effect of high-altitude cirrus clouds; and presumably other geoengineering responses may be imagined that do not constitute CDR or SRM either.[19] And finally, it is not clear how best to distinguish between CDR, SRM and the more commonly considered options for dealing with climate change: those of mitigation and adaptation.

This difficulty in categorising and situating geoengineering methods has recently been addressed in two papers, one by Clare Heyward and one by Olivier Boucher *et al.* Heyward proposes that rather than considering

---

[16] *Ibid.*, §2.2–3. [17] *Ibid.*, §3.3.
[18] The GGR terminology is suggested in O. Boucher *et al.*, 'Rethinking climate engineering categorization in the context of climate change mitigation and adaptation', *WIREs Climate Change*, 5 (2014), p. 31. In the Royal Society report where the partition of geoengineering techniques into SRM and CDR is suggested, it is acknowledged that methods to remove other GHGs might be sought (Shepherd *et al., Geoengineering the Climate*, p. 61).
[19] IPCC *Climate change 2013*, pp. 627–8. Cirrus thinning is 'not strictly a form of SRM', although it is often discussed alongside such techniques (*Ibid.*). This is because cirrus thinning would not reflect more of the sun's radiation back into space – as SRM is supposed to – but is instead designed to enable more outgoing, long-wave radiation (emitted from the Earth's surface) to escape by reducing the greenhouse effect of cirrus clouds.

geoengineering to be a third way to address climate change – alongside mitigation and adaptation – we may more usefully distinguish between five responses to this problem, and Boucher et al. offer an alternative five-way categorisation.[20] In what remains of this section, I draw on both accounts in order to provide a six-way typology of responses to climate change. I should note that Heyward presents her account not only in order to situate SRM and CDR with respect to other options for addressing climate change, but also to argue that the term 'geoengineering' should be abandoned because it is 'too broad [and] too vague' to be of use.[21] I will not heed this advice, instead drawing on Heyward's typology to offer a means by which we can attempt to determine whether a proposed response to climate change should be deemed to constitute geoengineering.

Following Heyward – and differentiating between responses to climate change 'by where they occur in the process between emitting GHGs and the loss of human well-being' – six categories of response to climate change that might be distinguished are:[22]

(A) the reduction of GHG emissions
(B) GGR
(C) SRM
(D) non-GGR/SRM geoengineering techniques
(E) adaptation
(F) rectification

Though they go about it in different ways, methods from categories (A) and (B) have in common that their ultimate aim is to reduce atmospheric concentrations of GHGs. Techniques in (C) are supposed to reflect more of the sun's radiation back into space. I introduce (D) as a category for geoengineering responses to climate change that *do not* fall into categories (B) or (C) – for example, cirrus thinning.[23] (E) encompasses adjustments designed to increase human resilience to climate impacts (as I will shortly explain, (E) may have some overlap with (C)). And finally – when other options fail to prevent important human interests being negatively impacted – measures from category (F) may be necessary to address the resulting loss and damage.

As the IPCC notes, whether a given activity constitutes geoengineering will depend on its 'magnitude, scale, and impact'.[24] I have defined category (D) such that it only contains techniques with sufficient magnitude and impact to constitute geoengineering. To decide which techniques from within (B) and (C) count as geoengineering, however, one must attempt to determine which of

---

[20] C. Heyward, 'Situating and abandoning geoengineering: a typology of five responses to dangerous climate change', *Political Science and Politics*, 46/1 (2013), 23–7, p. 25; Boucher et al., 'Rethinking climate engineering categorization', pp. 29–30.
[21] Heyward, 'Situating and abandoning geoengineering', p. 26. [22] *Ibid.*, p. 25. [23] See fn. 19.
[24] IPCC *Climate Change 2013*, p. 1449.

these methods would constitute a large-scale (regional or global) modification of the Earth system.²⁵

The distinction between large-scale and small-scale GGR is likely to be fuzzy and – when the effects of a method are hard to predict – it might not be clear in advance whether a given technique should be deemed to constitute geoengineering, mitigation, or both. Some might also debate whether a given GGR technique (e.g. large-scale industrial CCS) constitutes a *modification* of the Earth system or rather an attempt to reduce human modification of the climate. A GGR technique that *is* generally agreed to constitute geoengineering is ocean fertilisation. One that is not is the small-scale enhancement of terrestrial sinks.

Most SRM measures, on the other hand, are designed to have global effects and thus count as geoengineering by this definition – though not all. As Boucher *et al*. point out, some small-scale SRM methods (such as the introduction of more reflective roof and road surfaces and increased crop albedo) might more appropriately be considered adaptations. These measures have minimal trans-boundary effects, instead serving to cool the Earth's surface *locally*; with the potential benefit of preventing the 'urban heat island effect' and maintaining crop productivity.²⁶

As Christopher Preston suggests, it appears 'prudent' – given the broad range of methods that could be taken to constitute geoengineering – to evaluate the ethical aspects of each technique on an individual basis.²⁷ This separate analysis should help us to avoid getting bogged down in definitional disputes that distract from the more important, normative questions surrounding such proposals. Even if a particular GGR or SRM technique does not appear to constitute geoengineering, analogous ethical issues may arise – or alternative problems concerning the ability to access, control and profit from these inventions. Thus, as Heyward argues, it is important to focus on 'the specific features of proposed technologies' in order to assess their ethical and practical merit.²⁸

A number of geoengineering proposals are, however, united in their possession of three features that give rise to the particular normative issue on which I will focus. These techniques – including large-scale SRM, ocean fertilisation and cirrus thinning – have potentially *global effects*, which are expected to be *unevenly distributed* geographically, but are *currently uncertain*. In virtue of

---

²⁵ The difficulty in determining whether a given intervention would count as a *large-scale modification* of the Earth system helps to explain the controversy over which techniques should be regarded as geoengineering. Often the relevant scale is taken to be regional or global, but this doesn't resolve matters unless we know how large an area has to be before it counts as a *region*. Boucher *et al*. suggest an area around 300km x 300km (Boucher *et al*., 'Rethinking climate engineering categorization', p. 27), but any such stipulation will be ad hoc.
²⁶ Boucher *et al*., 'Rethinking climate engineering categorization', p. 26.
²⁷ C. J. Preston, 'Ethics and geoengineering: reviewing the moral issues raised by solar radiation management and carbon dioxide removal', *WIREs Climate Change*, 4 (2013), 23–37, p. 23.
²⁸ Heyward, 'Situating and abandoning geoengineering', p. 26.

the first two characteristics, such proposals raise a challenging problem of global justice: one that should be addressed through the creation of just and effective governing institutions. However, there is reason to think that international agreement on fair rules for the governance of geoengineering activities is more likely to be obtained in our current conditions of uncertainty regarding the global distribution of geoengineering impacts. When geoengineering proposals possess these three features, we therefore have reason to establish sound institutions of governance *before* attempting to reduce such uncertainty through further scientific investigation.

I should note, before continuing, that it is unclear how successful scientific investigation can actually prove in providing information about the probable impacts of geoengineering interventions. Some reject further research into geoengineering because they believe it is *impossible* to acquire such knowledge: Gerd Winter, for example, suggests that we might be in a state of 'unavoidable ignorance' regarding the safety of SRM techniques, and thus that further research into such proposals is pointless at best.[29] However, even if our ignorance regarding the distribution of geoengineering impacts is inescapable, the argument that follows remains applicable insofar as scientific studies have the potential to reduce the *subjective* uncertainty of parties in a position to implement geoengineering techniques – by leading them to believe that certain interventions will be in their self-interest.

Ethical and political issues concerning geoengineering

As stated above, in the rest of this piece I will be focussing only on issues raised by geoengineering techniques with potentially *global effects*, which are expected to be *unevenly distributed* geographically, but are *currently uncertain*. An obvious worry about such techniques is that they could have severe side effects for ecosystems and human societies. Their actual impacts are hard to predict, however, because there is no test system on which large-scale modifications of the Earth system can be trialled.[30] Projections must instead be made with the help of computer simulations, utilising models that currently appear far from adequate for this purpose – especially when tasked with predicting impacts at the local or regional level.[31] Thus, as Steve Rayner

---

[29] G. Winter, 'Climate engineering and international law: last resort or the end of humanity?' *Review of European Community and International Environmental Law*, 20/3 (2011), 277–89, p. 289.

[30] See A. Robock, M. Bunzl, B. Kravitz and G. L. Stenchikov, 'A test for geoengineering?' *Science*, 327/5965 (2010), 530–1.

[31] P. J. Irvine, A. Ridgwell, and D. J. Lunt, 'Assessing the regional disparities in geoengineering impacts', *Geophysical Research Letters*, 37/18 (2010), p. 1; Shepherd *et al.*, *Geoengineering the Climate*, p. ix, 54. Past volcanic eruptions offer some evidence regarding the expected effects of stratospheric particle injection. However, as an analogue they are far from ideal because their effects 'might only last for a few years', whilst stratospheric particle injection would have to

notes, disputants on every side of the debate surrounding geoengineering for now appear to be united in acknowledging 'that we currently know very little'.[32]

SRM, in particular, 'would itself inevitably lead to climate change', because it would have 'regionally different climate impacts'.[33] The effects of SRM on local temperatures will be varied and such methods have the potential to disrupt the hydrological cycle (leading to changes in precipitation) and to disturb atmospheric and oceanic circulation patterns. Furthermore, SRM could lead regional climatic conditions to migrate away from their historical baseline states in disparate ways as time progresses, 'meaning that "optimal" SRM activities imply different things for different regions'.[34] Some of these changes in regional climates 'might lead to impacts that are more serious than the amount of climate change that is being offset'.[35] In the case of cloud whitening, for example, 'while some areas benefit from geoengineering, there are significant areas where the response could be very detrimental'.[36] There is also reason to think that stratospheric particle injection might 'disrupt the Asian and African summer monsoons', resulting in drought conditions that would place 'the food supply for billions of people' at risk.[37] Similar effects on the Indian and African monsoons could arise from desert surface albedo modification.[38]

Despite the prospect of such negative consequences, some believe that we may end up in a situation where geoengineering is expected to be less catastrophic than climate change itself.[39] It might appear that in such circumstances, any ethical reasons against geoengineering would be overridden – and

take place 'for decades or centuries' (*Ibid.*, p. 29). A 'proper assessment' of the effectiveness of ocean fertilisation poses similar challenges, requiring 'consideration of the entire ocean carbon system, and the use of ocean carbon models' (*Ibid.*, p. 17).

[32] S. Rayner, 'To know or not to know? A note on ignorance as a rhetorical resource in geoengineering debates', *Climate Geoengineering Governance Working Paper Series*, 010 (2014), p. 16.

[33] D. J. Lunt et al., '"Sunshade world": a fully coupled GCM evaluation of the climatic impacts of geoengineering', *Geophysical Research Letters*, 35/12 (2008), p. 1; IPCC, *Climate Change 2013*, p. 551.

[34] K. L. Ricke, M. G. Morgan and M. R. Allen, 'Regional climate response to solar-radiation management', *Nature Geoscience*, 3 (2010), 537–41, p. 540. As Ricke et al. note, this suggests that 'as our understanding improves, serious issues of regionally diverse impacts and inter-regional equity may further complicate what is already a very challenging problem in risk management and governance'. In this chapter I defend a similar conclusion.

[35] M. MacCracken, 'Geoengineering: worthy of cautious evaluation?' *Climatic Change*, 77 (2006), 235–43, p. 237.

[36] A. Jones, J. Haywood, and O. Boucher, 'Climate impacts of geoengineering marine stratocumulus clouds', *Journal of Geophysical Research*, 114/D10 (2009).

[37] A. Robock, L. Oman, and G. L. Stenchikov, 'Regional climate responses to geoengineering with tropical and arctic SO2 injections', *Journal of Geophysical Research*, 113/D16 (2008).

[38] P. J. Irvine, A. Ridgwell and D. J. Lunt, 'Climatic effects of surface albedo geoengineering', *Journal of Geophysical Research*, 116/D24 (2011).

[39] K. Caldeira and D. W. Keith, 'The need for climate engineering research', *Issues in Science and Technology*, 27 (2010), 57–62. See also Shepherd et al., *Geoengineering the Climate*.

thus that geoengineering should be retained as some sort of insurance policy should such a tragic situation arise. Ken Caldeira and David Keith suggest that in order to make the best of this prudent back-up option, research into geoengineering must take place now so that such techniques are 'ready at hand (or at least as ready as possible)' *should they ever become necessary*. In fact, they argue, such investigation could *prevent* dangerous geoengineering from being implemented – or falsely assumed to provide a safety net – because *if* proposals are 'unworkable' or pose unacceptable risks, 'the sooner scientists know this, the faster they can take these options off the table'. Thus, whether the aim is to develop geoengineering as an insurance policy in case of a future tragic situation, or to prevent bad proposals from being deployed, it might appear that 'the stakes are simply too high for us to think that ignorance is a good policy'.[40]

Discussing the appropriate form of a potential research and development program for geoengineering, Caldeira and Keith concede that some forms of investigation – field testing, for example – are particularly controversial. Further lab-based research into the potential impacts of geoengineering, on the other hand, they claim to be a 'no-brainer', supported by many 'within and beyond the scientific community'.[41] In describing lab-based geoengineering research as a 'no-brainer', Caldeira and Keith seem to imply that seeking to reduce our uncertainty about these interventions through such investigation is morally unproblematic. Ralph Cicerone goes even further than this, defending lab-based geoengineering research by arguing that 'freedom of inquiry itself has moral value'; and Dale Jamieson, despite being concerned about the ethical issues raised by geoengineering, similarly suggests that 'the case for research in almost any field seems obvious and unassailable. It is better to know more than less'.[42]

I am not going to take a position, here, on whether the implementation of geoengineering could ever be morally acceptable. Instead, I will argue that *even if* some form of geoengineering may prove morally acceptable in a tragic situation; in our current circumstances, further lab-based research into the impacts of such techniques could have morally problematic outcomes. As Gardiner notes, superior knowledge doesn't necessarily help you to make the 'right' decision in a normative sense – it may instead help you to act unethically or unjustly. The consequences of such behaviour in the context of geoengineering could be extremely grave and therefore, I will argue, there is reason to think that maintaining our uncertainty about the impacts of geoengineering might be

---

[40] Caldeira and Keith, 'The need for climate engineering research', p. 62. See also D. W. Keith, E. Parson and M. G. Morgan, 'Research on global sun block needed now', *Nature*, 463 (2010), 426–7.
[41] Caldeira and Keith, 'The need for climate engineering research', p. 61.
[42] R. J. Cicerone, 'Geoengineering: encouraging research and overseeing implementation', *Climatic Change* 77/3 (2006), 221–6, p. 224; D. Jamieson, 'Ethics and intentional climate change', *Climatic Change*, 33 (1996), 331–2, p. 333.

wise for the time being.⁴³ I will explain why uncertainty may currently be the best policy by first discussing the appropriate form of governance for geoengineering activities, and then explaining how uncertainty could help to promote international agreement on such governance.

The governance of geoengineering

Even if there could be a future state of affairs in which geoengineering would be the lesser of two evils – and thus perhaps the thing to do – we can assume that such a tragic situation would be quite exceptional. There is a much broader range of circumstances in which geoengineering deployment would be both ethically and politically unacceptable: for example, because the situation does not warrant such action, the risks are too great, or the decision to implement lacks legitimacy.

We are therefore in need of some form of international regulatory regime, designed to ensure that geoengineering activities are governed in accordance with fair rules or collective decision-making procedures. However, though efforts are already being made to address the question of global governance for geoengineering⁴⁴ – and some existing international rules have applicability in this arena⁴⁵ – binding international agreement on regulative principles is yet to materialise. It therefore remains the case that 'no international treaties or institutions [have] a sufficiently broad mandate to regulate the broad range of possible geoengineering activities', presenting 'a risk that methods could be applied by individual nation states, corporations or one or more wealthy individuals, without concern for their transboundary implications'.⁴⁶

---

⁴³ I thus adopt a position that Rayner claims is common to opponents of geoengineering research: that 'it is precisely because the stakes are so high that ignorance is . . . a good policy' (Rayner, 'To know or not to know?', p. 13). Cf. Caldeira and Keith, 'The need for climate engineering research', p. 62.

⁴⁴ See, for example, the Oxford Principles (S. Rayner *et al.* 'The Oxford Principles', *Climatic Change*, 121 (2013), 499–512); the Asilomar Conference Recommendations (Asilomar Scientific Organizing Committee, *The Asilomar Conference Recommendations on Principles for Research into Climate Engineering Techniques* (Washington, DC: Climate Institute, 2010)); the Solar Radiation Management Governance Initiative (www.srmgi.org); and the Bipartisan Policy Centre Report (bipartisanpolicy.org/library/report/task-force-climate-remediation-research, last accessed 9 June 2014).

⁴⁵ For example, The Convention on Biological Diversity (www.cbd.int/climate/geoengineering/default.shtml, accessed 28 May 2014); the Convention on Environmental Modification (www.un-documents.net/enmod.htm, last accessed 28 May 2014); and the London Convention and Protocol, imo.org/OurWork/Environment/LCLP/Pages/default.aspx, last accessed 28 May 2014). A problem with most of the legal norms that have relevance in this area is that they 'impose little meaningful constraint on geoengineering activities because they are very general and leave states with a huge amount of discretion in deciding what to do' (D. Bodansky, 'The who, what, and wherefore of geoengineering governance', *Climatic Change*, 121 (2013), 539–51, p. 542).

⁴⁶ Shepherd *et al., Geoengineering the Climate*, p. 60. See also E. A. Parson and L. N. Ernst, 'International governance of climate engineering', *Theoretical Inquiries in Law*, 14 (2013), 307–38, 320ff.

The importance of fair and effective geoengineering governance is widely accepted, though defenders of lab-based research often suggest that the need for such institutions only arises once field-testing or deployment are considered. The report from the 2010 Asilomar Conference on geoengineering, for example, recommends that new governance mechanisms be created for 'large-scale climate engineering research activities that have the potential or intent to significantly modify the environment or affect society' – though not for lab-based research, since this poses 'no novel risks or challenges'.[47] Caldeira and Keith similarly acknowledge the importance of governance once field-testing is proposed, but hold that lab-based research should be pursued now despite the absence of such institutions.[48] And though Cicerone is particularly resolute concerning the need for governance, suggesting a 'moratorium on large-scale field manipulations' until 'acceptable agreements' on ethical and legal issues have been made and a governing body established to oversee such activities, he argues that scientists should support lab-based research in the interim.[49] The idea, presumably, is that lab-based research should proceed in tandem with the design and establishment of appropriate governing institutions, with the moratorium on field-testing lifted if and when both are deemed to have achieved adequate results.

The design of governing institutions raises obvious difficulties in that while geoengineering could impact all human beings, many of these methods might be implemented unilaterally – by a single state or even a sufficiently wealthy individual or corporation. Acute problems of consent, control, authority, legitimacy, monitoring and enforcement therefore arise. The design of such institutions also raises extremely challenging questions of global justice. As explained above, any given geoengineering intervention is likely to have varied local impacts and will create winners and losers. Even if geoengineering appears likely to reduce harm or death, then: it is quite possible that different people will be harmed or killed as a result of such interventions, creating a profound problem of global distributive justice.

Plausibly, just rules to govern geoengineering activities would, at a minimum, show equal consideration to the interests of all human beings (rather than being designed to serve the interests of some subset of individuals such as, for example, the citizens of a particular state); would provide strong protection for those vulnerable to harm via the testing or implementation of such techniques; and would be the subject of some form of agreement by all affected parties or their

---

[47] Asilomar Scientific Organizing Committee, *The Asilomar Conference Recommendations on Principles for Research into Climate Engineering Techniques* (Washington, DC: Climate Institute, 2010), p. 18. See also D. R. Morrow, R. E. Kopp and M. Oppenheimer, 'Toward ethical norms and institutions for climate engineering research', *Environmental Research Letters*, 4 (2009), p. 2.

[48] Caldeira and Keith, 'The need for climate engineering research', pp. 61–2.

[49] Cicerone, 'Geoengineering', pp. 224–5.

representatives. I will not devote much time to the important question of whose agreement would be necessary to render these institutions legitimate, except to note that while it is of practical importance that state representatives assent to such global governance,[50] it is clear that other parties should also be involved in the process of institutional design.[51] I shall instead focus on how likely it is that the relevant parties will be able to agree on equitable rules of geoengineering governance, which offer adequate protection for the vulnerable.

At first glance, the chances of such agreement would appear slim given the presumed involvement of state representatives; whose primary objective will likely be that of ensuring that governing principles advance their own national interests, and whose track record in addressing climate change has thus far been lacking, to say the least.[52] In the subsequent sections of this chapter, I argue that our current circumstances of scientific uncertainty regarding the local and regional impacts of geoengineering interventions could actually have normative value – in the sense that they may be conducive to agreement on fair rules for the governance of activities within this sphere. This suggests that if we want to ensure that just and effective governance is in place prior to any testing or implementation of geoengineering, we may also have to insist – contra the defenders of lab-based research discussed above – that such institutions are created before further investigation into the likely distribution of geoengineering impacts.

The normative value of uncertainty

Though the suggestion that uncertainty can have normative value might appear somewhat unfamiliar, a link between uncertainty and fairness can be found in well-known theories from philosophy, economics and game theory.[53] One of the most prominent uses of this idea in philosophy is found in the work of John Rawls. In *A Theory of Justice*, Rawls attempted to formulate fair principles to

---

[50] As Bodansky points out, given the lack of strong executive powers at the global level, international rules and decision-making procedures 'depend primarily on self-compliance by states' (Bodansky, 'The who, what, and wherefore of geoengineering governance', p. 544).

[51] Indigenous peoples being an important example (see K. P. Whyte, 'Now this! Indigenous sovereignty, political obliviousness and governance models for SRM research', *Ethics, Policy and Environment*, 15 (2012), 172–87).

[52] Concerns along these lines are raised by Gardiner, who worries that 'current political inertia on climate change' could well result in part from 'a resistance to the kinds of norms of global justice and community that dealing with the problem might suggest' – a problem likely to 'infect the search for legitimate geoengineering governance' (S. M. Gardiner, 'Some early ethics of geoengineering the climate: a commentary on the values of the Royal Society report', *Environmental Values*, 20/2 (May 2011) 163–88, p. 168 and p. 171).

[53] Binmore goes so far as to suggest that uncertainty about the future may help to explain the very origins of our fairness norms (K. Binmore, *Natural Justice* (Oxford: Oxford University Press, 2005), ch. 9).

govern the institutions of a single society using a device called the original position. The original position is a hypothetical situation of choice under uncertainty, where parties representing individual citizens seek agreement on the principles by which their societal institutions will be governed. The epistemic restrictions to which the parties are subjected (termed the 'veil of ignorance') conceal all information regarding their personal identity – denying them any knowledge even of probabilities that their individual circumstances will be characterised by certain features. The parties thus do not know how the principles agreed to will affect them personally, and in particular cannot rule out that they will prove to be among the worst-off in the resulting distribution of costs and benefits.[54]

This total uncertainty serves two purposes: first, it aligns the interests of the parties, because in ignorance of their own circumstances they must choose principles on the basis of general facts and concerns alone; and second, it ensures that the principles consented to are fair, because the parties lack any information that they could use to bias the agreement in their own favour. In having to seek agreement on institutions that will best serve their interests *whatever* position in society they may turn out to occupy, the parties are supposed to end up selecting principles that serve everybody's interests equally. In particular, Rawls claims, one can imagine that parties in this situation might well decide that the sensible thing to do is *maximin*: to rank the principles that they can choose between 'by their worst possible outcomes', and select the option with the worst outcome that is 'superior to the worst outcomes of the others'.[55] In this sense, the constraints of uncertainty direct the pursuit of self-interest towards agreement on fair principles:[56] principles that take everybody's interests into account, and protect the least well-off in society by providing them with a 'satisfactory minimum'.[57]

Though such extreme epistemic restrictions are absent in real-world decision-making, one might think that uncertainty deriving from other sources could have similar effects. For example, James Buchanan and Geoffrey Brennan – in their discussion of the reasons that individuals have for preferring certain political rules – argue that as the *permanence* of such rules increases, a fairness-promoting form of uncertainty may be brought about.[58] The longer a given political rule is intended to be in place, the greater the uncertainty faced by individuals that will be subject to it; in the sense that it becomes harder to predict how that rule will affect them personally, the further into the future they must look. Even citizens in full knowledge of their current circumstances are

---

[54] J. Rawls, *A Theory of Justice: Revised Edition* (Oxford University Press, 1999), pp. 118–19.
[55] *Ibid.*, p. 26. [56] *Ibid.*, pp: 127–9. [57] *Ibid.*, p. 135.
[58] G. Brennan and J. M. Buchanan, *The Reason of Rules: Constitutional Political Economy* (Library of Economics and Liberty, 2000), §10.2.35–8 (www.econlib.org/library/Buchanan/buchCv10.html, accessed 21 May 2014).

thus placed in a situation of choice under uncertainty when they seek agreement with one another on *long-term* rules by which they will be collectively governed.

Buchanan and Brennan argue that this effect – which they refer to as the 'veil of uncertainty' – provides individuals with reasons of self-interest 'to concentrate on choice options that eliminate or minimise prospects for potentially disastrous results'. Parties selecting long-standing political rules by which they shall be collectively governed will therefore 'tend to agree on arrangements that might be called "fair" in the sense that patterns of outcomes generated under such arrangements will be broadly acceptable, regardless of where the participant might be located in such outcomes'.[59] Robert Goodin and John Dryzek suggest that a similar effect on the selection of political rules may be brought about by circumstances of flux and turmoil. When conditions are unpredictable (for example, in a situation of war), individuals face significant uncertainty not just about the distant future, but also the near future. It thus becomes harder for citizens to ascertain where they will end up in a societal distribution of costs and benefits, with the result that it is in everyone's interests for political institutions to provide the poor with a social safety net.[60]

Though the underlying uncertainty in each of these cases is different – relating to personal identity, the distant future or the near future – the reason to think that it might have normative value is the same. This underlying uncertainty creates uncertainty about where individuals will be situated in the distributions of costs and benefits resulting from different social and political arrangements, and *this* uncertainty about whether one will be a winner or a loser with respect to such institutions ensures that even self-interested parties have reason to prefer collective rules that possess elements of fairness. The mechanism by which this can take place is quite simple: when parties are denied information that would enable them to identify the rules that will best serve *their own* interests, they have more reason to agree on rules that serve *everybody's* interests – rules, for example, that maximise the position of the least well-off, result in broadly acceptable patterns of outcomes, or provide the vulnerable with a safety net.[61]

---

[59] *Ibid.*, §10.2.36–8.
[60] R. E. Goodin and J. Dryzek, 'Risk sharing and justice: the motivational foundations of the post-war welfare state', in R. Goodin and J. LeGrand (eds.), *Not Only the Poor: the Middle Classes and the Welfare State* (London: Allen and Unwin, 1987), pp. 37–73. Goodin and Dryzek here suggest that this phenomenon could help to explain the substantial growth of the welfare state that took place in a number of countries during and after the Second World War.
[61] Rawls' claim that it is rational for parties in the original position to maximin has been extensively criticised. See, for example, J. C. Harsanyi, 'Can the maximin principle serve as a basis for morality? A critique of John Rawls's theory', *The American Political Science Review*, 69 (1975), 594–606. Whether or not such criticism succeeds, it may nevertheless be the case that less-than-ideally-rational individuals would reason this way, in relevantly similar real-world situations of choice under uncertainty.

Economic theorists interested in climate change have started to consider the possibility that the uncertainty we face in dealing with this problem could have such normative value, using models to explore how 'uncertainty and the prospect of learning' affect the incentives for countries to join an international environmental agreement.[62] Though conclusions are only tentative, some of the models so far considered have suggested that uncertainty regarding how the costs and benefits of environmental policies will be distributed could increase the likelihood of cooperation – and that removing such uncertainty might reduce the total amount of welfare that can be obtained by forming an international environmental agreement.[63] As Seong-Lin Na and Hyun Song Shin put it, 'cooperation is more difficult to achieve when the likely winners and losers are known when negotiation takes place'.[64] There is thus reason to think that in certain situations, information about environmental impacts could have 'negative social value', because by revealing the winners and losers of different environmental policies, it undermines incentives for cooperation.[65]

To summarise this section, then: there is reason to think that when a number of parties seek to establish the arrangements by which they will be collectively governed, uncertainty about how the costs and benefits of potential rules will be distributed could be conducive to agreement on fair institutions. This is because such uncertainty restricts information that parties can use to bias agreements in their own favour. In this situation, self-interested agents are given reason to adopt the common goal of securing their own welfare via institutions that will serve everybody's interests – and, in particular, protect the interests of the most vulnerable. If uncertainty about how the costs and benefits of different environmental policies will be distributed possesses such normative value, however, then research that serves to reduce such uncertainty could have morally problematic outcomes.

Why (effective) governance should come before (certain kinds of) research

The analogy between the examples of decision making under uncertainty discussed above and the context of institutional choice for geoengineering governance should by now be obvious. In order to ensure that geoengineering research and potential implementation will take the interests of all human

---

[62] C. D. Kolstad and A. Ulph, 'Uncertainty, learning and heterogeneity in international environmental agreements', *Environmental and Resource Economics*, 50/3 (2011), 389–403, p. 391.

[63] C. D. Kolstad, 'Piercing the veil of uncertainty in transboundary pollution agreements', *Environmental and Resource Economics*, 31 (2005), 21–34; S. Na and H. S. Shin, 'International environmental agreements under uncertainty', *Oxford Economic Papers*, 50 (1998), 173–85; Kolstad and Ulph, 'Uncertainty, learning and heterogeneity', p. 389.

[64] Na and Shin, 'International environmental agreements under uncertainty', p. 176.

[65] Kolstad and Ulph, 'Uncertainty, learning and heterogeneity', p. 401.

beings into account, we need states and other relevant parties to agree on fair rules for the governance of their activities within this sphere. At present, those charged with designing and selecting such rules would have to do so in circumstances of significant uncertainty regarding their position in the distribution of costs and benefits that could result. The reasoning of the previous section suggests that such uncertainty may actually have normative value insofar as it could facilitate agreement on fair rules of governance – and thus that we should be wary of reducing such uncertainty before the necessary institutions have been put in place.

This uncertainty about where the costs and benefits may fall, with respect to the rules that could potentially be used to regulate geoengineering activities, derives from a number of sources. For one thing, our world is transforming quickly as a result of climate change and technological progress, which will make it difficult for parties to decide how rules for the governance of geoengineering will affect their interests (or their national interests) over time. An even more significant underlying uncertainty, however, is our extreme scientific uncertainty concerning what the local and regional impacts of any given geoengineering intervention will be. Such uncertainty makes it much harder to determine what form of geoengineering governance will prove to be in any group or individual's best interests, thus creating difficulty for parties seeking to ensure that the benefits of regulatory decision-making are skewed in their own favour.

In particular, it is currently impossible for any parties to rule out the possibility that geoengineering testing or deployment would subject them to severe negative impacts – ones to which they cannot adapt, and for which there is no possibility of rectification. This means that all parties have some incentive to ensure that regulation offers effective protection to those vulnerable to harm. Given the potentially catastrophic outcomes that could result if mechanisms are inadequate to prevent others from implementing geoengineering techniques, parties may even be motivated to maximin; that is, to seek agreement on rules that are likely to maximise the position of the most disadvantaged (where this will presumably include those who stand to be worst off as a result of climate change or potential geoengineering impacts).[66] Alternatively, international actors may at least seek to ensure that geoengineering research and potential deployment is regulated in a way that is designed to promote broadly acceptable outcomes, with protections in place for the most vulnerable.

The claim that further lab-based research into geoengineering is a 'no-brainer' is therefore more problematic than it appears, despite the broad

---

[66] Rawls suggests that maximin decision making is most plausible in a situation of choice under uncertainty involving 'grave risks' (Rawls, *A Theory of Justice*, p. 134).

support that such inquiry receives.[67] By reducing uncertainty about regional and local impacts – thereby helping to identify the winners and losers of potential geoengineering policies and interventions – this research could undermine international cooperation by revealing divergent interests with respect to the regulation of geoengineering. Whether or not freedom of inquiry is good in itself, then; it is important to consider whether pursuing certain avenues of research could have consequences that are bad enough – morally speaking – to outweigh any positive normative value that such inquiry may possess.

If our current circumstances really do instantiate one of those 'rare moments of deep and widespread uncertainty' that can serve to promote moral behaviour, then it is important to capitalise on this opportunity.[68] One simple way to do this would be to temporarily extend Cicerone's proposed moratorium on scientific research to include lab-based studies into the regional distribution of geoengineering impacts – with the moratorium to continue until fair and effective institutions of global governance were established. This is not an unreasonable demand: if the necessary parties cannot prove that they will be able to govern geoengineering in an ethically acceptable manner, then it is risky to engage in research that might lead certain international actors to believe that they would benefit from a particular use of such technology (despite corresponding costs to others). Furthermore, this demand is particularly hard to reject with respect to techniques that are viewed as a lesser evil, to be used as an option of last resort. Such interventions are not expected to be necessary in the near future, so halting research until adequate governance mechanisms have been established should not lead to irrecoverable delays.

As Gardiner has argued, the problem of climate change presents the current generation with a moral challenge that is perhaps unprecedented in its global scope, inter-generational nature and the inadequacy of our current normative theories for dealing with many of the questions that it raises. In these testing circumstances, we ought to be on the lookout for signs of moral corruption; for signs, that is, that those in a position of power (the affluent and current generations) are seeking to obscure their attempts to shift the costs of climate change onto those who are relatively weak (the poor and future generations).[69] Gardiner's warning about moral corruption should lead us to be vigilant if

---

[67] The Royal Society, for example, states that the development and use of new models to investigate geoengineering should be a high research priority, and identifies the 'spatial heterogeneity' of SRM impacts as a particular area in which further study should be conducted (Shepherd et al., Geoengineering the Climate, pp. 52–4).

[68] Goodin and Dryzek, 'Risk sharing and justice', p. 66; Rayner similarly suggests that 'once research gets under way (if it does), then ... debates over the science will displace the debate over the values', meaning that we currently have 'a moment of clarity' that we should attempt to capture and use to stimulate discussion about the kind of world that we want to inhabit (Rayner, 'To know or not to know?', pp. 15–16).

[69] Gardiner, A Perfect Moral Storm, pp. 7–8.

certain parties appear to be trying to delay or undermine the establishment of strong regulatory institutions for geoengineering governance. The argument presented in this chapter suggests that those attempting such obstruction could be motivated by the desire to find out more about the likely impacts of geoengineering before international rules are put in place, so that they can seek to bias any governing agreements in their own favour; or, even, retain the option of geoengineering deployment in pursuit of their own benefit, unhindered by strong regulatory principles.[70]

Conclusion

Geoengineering is generally conceded to be an ethically problematic response to the problem of climate change, even by those who argue that we must retain this alternative as an option of last resort and that scientists should therefore continue to engage in geoengineering research so that we can be prepared for this eventuality. Fair and effective governance is also widely accepted to be essential prior to any testing or implementation of these techniques, in order to ensure that such actions are legitimate and morally acceptable.

In this chapter I have given reason to think that as uncertainty about the international distribution of geoengineering impacts is reduced, our chances of establishing fair and effective institutions of global governance will diminish along with it. Uncertainty about the impacts of geoengineering methods currently makes it hard to assess who the winners and losers will be, with respect to the different forms of regulation that might be adopted. This creates a barrier to the exercise of self-interest and gives all parties to any governance agreement a reason to be concerned with the plight of those who would bear the costs of weak or inequitable regulation – because for all they know, they could prove to be among those most vulnerable to the potentially catastrophic impacts of geoengineering. Thus, even lab-based research into the impacts of geoengineering interventions could be more ethically problematic than scientists recognise. If such research leads a particular state or other party to conclude that they are a likely winner with respect to a given geoengineering intervention, any incentive to sign up to fair and effective governing institutions will be severely diminished – just as their temptation to pursue this option will become dangerously enhanced.

---

[70] See, for example, the argument given in L. Lane, 'Climate engineering and international law: what is in the national interest?' *Proceedings of the Annual Meeting (American Society of International Law)*, 105 (2011), 525–8. Lane here defends further geoengineering research on the basis that it will enable the US to identify what policies are in its national interest; and rejects any governance or international law *at all* so as to preserve 'America's freedom of action' within this sphere (*Ibid.*, pp. 527–8).

Advocates of geoengineering are often accused of arrogance for supposing that human beings can control the climate, but similarly worrying appears to be the assumption that scientists can control what is done with the results of their research. Whether or not geoengineering could turn out to be the lesser of two evils in a certain, tragic future scenario – it is not clear why one would think that this is the only situation in which international actors might decide to implement such techniques. Caldeira and Keith are correct that research into geoengineering may ultimately suggest that all such proposals pose unacceptable risks to sections of human society. What may prove hubristic, or simply naive, is the assumption that scientists will be able to take these unacceptably risky options off the table, should their research imply that such techniques would simultaneously benefit parties with the capacity to geoengineer – parties who may decide that the potential benefits to themselves render the risks to others acceptable after all.

3 Climate justice and territorial rights

Chris Armstrong

Introduction

When we engage in activities that cause greenhouse gases (GHGs) to be emitted, we thereby consume some of the capacity of the world's 'carbon sinks' – and their capacity, we now understand, is limited.[1] The sequestering capacities of these sinks function as a 'collective' good: one which individuals cannot be practically excluded from consuming, but also one where each act of consumption leaves less for others now and in the future. Political theorists have, accordingly, devoted considerable attention to the question of just how permissions to use up sink capacity should be distributed.[2]

One very well-known answer to that question is the equal per capita principle, which would simply divide extant sink capacity into equal slices, setting for each person an identical emissions budget (on many proposals, unused emissions space might then be traded with others). The normative appeal of that principle is often thought to be self-evident. Surely, the thought goes, our entitlements to make use of carbon sinks must be symmetrical? After all, none of us put those sinks there and no-one, therefore, appears to have any special claim over them. In his discussion of rights over atmospheric sinks, Peter Singer has suggested that 'If we begin by asking "Why should anyone have a greater claim to part of the global atmospheric sink than any other?" then the first, and simplest response is: "No reason at all"'.[3] If it were adopted by a global agreement on climate mitigation, that principle would have radical

---

[1] Just how large we consider their capacity to be will hinge on the degree of climate change we believe is tolerable. Once we fix on a target – such as keeping global mean temperature rises under 2°C – we can then estimate a cumulative total for the greenhouse gases that can safely be emitted by human activities. The 'Trillionth Tonne' project, for instance, estimates that cumulative total at one trillion tonnes of carbon dioxide (CO_2), measured since 1750, and calculates that at current consumption levels we will have exhausted that capacity within thirty years. See www.trillionthtonne.org.

[2] Thanks to Megan Blomfield for a helpful discussion of some of the issues discussed in this chapter.

[3] P. Singer, *One World: the Ethics of Globalization*, 2nd edn (New Haven: Nota Bene Press, 2004), p. 35.

effects. At present inhabitants of industrialised countries consume a share of the capacity of the world's carbon sinks far in excess of what would be sustainable if others acted in the same way. The principle would ask them to severely contract their emissions, and allow inhabitants of developing countries to eventually 'converge' upwards on the per capita emissions benchmark.

But on closer inspection the principle of equal per capita emissions (hereafter EPC) loses much of its appeal. People consume a wide range of goods, and most theories of distributive justice express concern with holdings in a fairly broad set of them. Why then insist on equal shares of *one* good in particular? To be sure reasons can sometimes be provided for insisting on equal shares of particular goods, and for seeing shares of different goods accordingly as at least partly incommensurable. The principle of 'one person, one vote' is thought, by those committed to democracy, not to bend in light of facts about wealth or educational access. We do not simply give the poor or the educationally deprived more votes – at least not if we are committed to seeing ourselves as equal citizens. But such reasons do not appear to apply in the case of the world's GHG-sequestering capacity.[4] Given that they do not, insisting on equal emissions rights would be myopic even from an egalitarian point of view. Even if we restrict our attention to climate justice – rather than justice more broadly – egalitarians should prefer an account which considers a range of ecological 'credits' and 'debits', and calls for equality across this broad bundle of benefits and burdens. Some countries, to give an example, may produce massive quantities of harmful pollutants whereas others may go to great lengths to avoid doing so. Some countries – but not others – may spend large sums of money subsidising the spread of green technologies in the developing world. Some countries, moreover, might already be better off in other dimensions relevant to justice. All of these facts should matter when we come to allocate emissions rights. From that perspective, while EPC might still be a big improvement on the status quo, it begins to look far from ideal.

There may, though, be a second reason for doubting the appeal of EPC, and it will be my focus in this chapter. As I just suggested, the argument for EPC usually leans on an assumption that no-one has created, or is responsible for the existence of, the Earth's GHG-sequestering capacity.[5] But perhaps this is too

---

[4] For a full and persuasive argument to that effect, see S. Caney, 'Just emissions', *Philosophy and Public Affairs* 40/4 (2012), 255–300. See also C. Armstrong, 'Natural resources: the demands of equality', *Journal of Social Philosophy*, 44/4 (2013), 331–47.

[5] See Singer, *One World*. For further defence of equal emissions on the basis of a view about sinks as global commons, see for instance P. Baer, 'Equity, greenhouse gas emissions, and global common resources', in S. Schneider *et al.* (eds.), *Climate Change Policy: a Survey* (Washington DC: Island Press, 2002), pp. 393–408; S. Vanderheiden, *Atmospheric Justice* (Oxford University Press, 2008). For a view that common ownership of the atmosphere should be seen within the context of common ownership of the Earth more broadly – precisely because no-one created either – see M. Risse, *On Global Justice* (Princeton University Press, 2012), ch. 10.

quick. It may well be true, as Singer suggests, that no-one created the *atmosphere's* sequestering capacity. We might say the same about the oceans. But it is less certain that we can say the same thing for *all* carbon sinks. Perhaps agents are (at least sometimes) responsible for the sequestering capacity of sinks currently under their territorial control, for instance. 'Terrestrial' carbon sinks include, most famously, tropical rainforests – the so-called 'lungs of the world' – but also the soil, lakes and other vegetation contained in particular geographical areas. These terrestrial sinks, significantly, are located within the borders of states which might be thought to have territorial rights over them, and it *might* be thought that citizens of those states have special claims over their sequestering capacity.

If the argument holds, one possible implication is that the capacity of such sinks ought to be considered, within any treaty on climate justice, as a 'credit' on the part of the states within which they are located. Perhaps the greenhouse gases absorbed by a state's rainforests ought to be deducted from its notional emissions total (or, to put it the other way, added to its emissions quota). If good arguments can be made to that effect, then it looks as though a serious engagement with questions about territorial rights is necessary and that it looks likely to complicate what we must say about climate justice. Theorists of climate justice could then not avoid taking a stance on thorny questions about just who should enjoy territorial rights over which parts of the Earth. But does the argument hold? This chapter critically examines the arguments that might be made in favour of treating sink capacity as a 'credit' on the part of states with terrestrial sinks within their borders, and in fact suggests that they do not deliver on the conclusion that states with terrestrial sinks should be granted extra emissions entitlements. The final section, though, considers a quite different argument which suggests that we may nevertheless have duties to make transfers to states with important carbon sinks. Even if we reject the idea that terrestrial sink capacity belongs to the states in which it resides, we can nevertheless have duties of justice to offset the costs involved in protecting that capacity. We can embrace payments for sink protection, then, even at the same time as rejecting local sink ownership.

Who owns carbon sinks?

A young hardwood tree stands in a tropical forest surrounded by millions of others like it. As it grows, it absorbs carbon dioxide from the air (while also drawing water from the Earth, and fixing energy from the sun within its leaves). When it dies much of the carbon dioxide it contains will be released back into the atmosphere, with the remainder entering the soil. But in the meantime, it represents a tiny carbon sink. So just who, if anyone, is the rightful owner of its sink capacity? All of humankind? Members of the indigenous community who eke out a living in the forest nearby? Or the population of the state in which the forest lies, granted 'permanent sovereignty' over their natural resources under international law?

In an important recent piece, Megan Blomfield has examined some of the arguments which might be introduced to justify claims on the part of local communities. While she does not argue for *state ownership* of sink capacity (since she remains agnostic both on the claims of states as opposed to other communities, and on whether ownership of sink capacity is the best response to sound claims), her argument presents a useful opportunity to assess the basis and strength of any local claims. One objection to equal per capita emissions, she notes, is precisely that it ignores the fact that GHGs are absorbed, inter alia, by terrestrial sinks over which individual states are often thought to have rights.[6] Although the atmosphere may rightly be regarded as a global commons, perhaps terrestrial sinks should not. Respect for territorial rights may therefore produce at least a partial answer to the rhetorical question posed by Singer. If my state contains major *terrestrial* sinks, the answer might go, then perhaps I *should* be allowed to consume more of the Earth's sequestering capacity.

But what is the basis of local claims over terrestrial sinks? And how robust are those claims? That will depend on the robustness of the general case that can be made for states' control over natural resources. It seems to me that three arguments might be invoked in defence of communal resource claims, and Blomfield advances versions of all three in order to render plausible the view that local communities may have claims over the sequestering capacity of sinks contained within their territories.[7] But are the three arguments likely to deliver on the conclusion that locals have a claim over the capacity of terrestrial sinks?

Attachment

Consider, first, the argument from attachment. Here the thought is that the communities which currently control terrestrial sinks (such as rainforests) might be strongly attached to them, and indeed their very identities might be bound up in their control over or access to them.[8] If a community is strongly attached to a sink in this way – such that control over it or access to it is essential

---

[6] M. Blomfield, 'Global common resources and the just distribution of emissions shares', *Journal of Political Philosophy*, 21/3 (2013), 283–304.

[7] The three arguments I canvass here are roughly equivalent to what Lea Ypi calls arguments from attachment, from legitimacy and from acquisition. L. Ypi, 'Territorial rights and exclusion', *Philosophy Compass*, 8/3 (2013), 241–53. For a fuller discussion of these arguments and an examination of their general prospects of supporting permanent sovereignty in particular, see C. Armstrong, 'Against "permanent sovereignty" over natural resources', *Politics, Philosophy and Economics*, 14/2 (2015), 129–51.

[8] Blomfield, 'Global common resources', pp. 294–6.

to that community's sense of self – then perhaps this gives us a reason to treat the sequestering capacity of that sink as something that properly belongs to the community in question.[9] I have suggested elsewhere that theorists of justice both can and should take such attachments seriously. If we have reasons for caring about life-plans which play a central part in people's lives – and surely a wide range of theories of justice can marshal such reasons – then we derive reasons for caring about people's secure access to the supports for those plans, which in some cases will include specific natural resource tokens. Even if a concern for such life-plans does not establish that people with significant attachments to resources are entitled to *more* than others, it can establish a *pro tanto* reason for permitting such people to retain access to the *particular* resources they happen to be attached to.[10]

But when we invoke arguments from attachment we need to carefully specify not only which resources are crucial to important life-plans, but also which rights over those resources those life-plans are genuinely dependent on. Most importantly, we should not assume that wherever attachment over an external object is at stake nothing other than full rights of 'liberal ownership' are capable of supporting the relevant plans. Very often they will turn out to demand something considerably more limited than that. In many cases, attachment appears to require that the agent so attached should have the right to securely access a resource, and this requires that we prevent others from degrading or destroying it or placing it beyond their reach. It is far from obvious, though, that respecting attachment requires that we grant the attached party ownership over any and all benefits flowing from the resource in question. For instance, imagine that I am attached to a river near my home. One of my most central life-plans is to continue to fish from it, bathe in it and drink from it. Protecting this ability requires continued access and, in this case, some rights to withdraw fish and water (which rights might also, plausibly, be held by many others). It does not appear to require that when others obtain benefits from that river they pay me for doing so. Their plans and mine are entirely compatible without any such side-payments. Respecting my attachment to the river does not justify granting me a monopoly over its benefits or a full set of resource rights.

---

[9] For an argument from the creation of 'symbolic value' to national rights over land and resources, see D. Miller, 'Territorial rights: concept and justification', *Political Studies*, 60/2 (2012), 252–68.

[10] C. Armstrong, 'Justice and attachment to natural resources', *Journal of Political Philosophy*, 22/1 (2014), 48–65. Avery Kolers has presented a more demanding account of what follows from what I am calling attachment to resources; see A. Kolers, 'Justice, territory and natural resources', *Political Studies*, 60/2 (2012), 269–86. I reject his account in C. Armstrong, 'Resources, rights and global justice: a response to Kolers', *Political Studies*, 62/1 (2014), 216–22.

Let us return to the case of rainforests with this caution in mind. We can again readily perceive that attachment to a rainforest – or the on-going and central project of being a forest-dweller, say – is going to be thwarted when we bar people from a forest, or when we destroy it. But of course no-one in debates about climate justice is suggesting that we should exclude locals from the rainforests, or that we should tear them down. Quite the opposite: outsiders would prefer rainforests to continue to exist, to the extent that there is a live debate about whether locals should actually be *paid* not to tear down their forests (see the section 'Sharing the costs of protecting carbon sinks'). But that is entirely distinct from the question of who has the right to benefit from their sink capacity. Here it is simply not clear that attachment-based claims over geographical sites like rainforests are thwarted when we treat their capacity as shared assets rather than private ones. We can respect the rights of forest-dwellers to dwell in forests, or of forest peoples to *see* themselves as, and to engage in the customary practices of, forest peoples, without granting forest-dwellers ownership over the sink capacity of the forests they dwell in when we come to draw up a global agreement on climate justice. Their life-plans do not appear to depend upon whether the greenhouse gases absorbed by the forest are French, Chinese or Tunisian in origin.[11] Respecting the attachment-based special claims of forest-dwelling peoples appears compatible, then, with a wide range of views on how we should allocate sink capacity.

Self-determination

Consider, second, the argument from self-determination. This argument will attempt to show that, since control over natural resources is an important component of the self-determination of a political community, to deny that the sequestering capacity of a rainforest belongs to the country in which it is contained is to deny that country its right of self-determination. Accordingly, the (steep) challenge faced by self-determination-based arguments is to show how control over any *particular* resource or set of resources is indeed indispensable for self-determination. Any view which stipulates that full control over all of the resources which happen to fall within a state's territory is essential for its self-determination will, it seems to me, stretch credulity, and moreover will be unappealing as a conception of self-determination to anyone concerned with global distributive injustices. After all, some countries possess resources greatly in excess of anything they might feasibly use themselves, whereas other countries – which presumably have just the same right to be

---

[11] Might the claim be that they cannot pursue those plans without having exclusive rights to use that capacity themselves? It is hard to see how such a claim could succeed. We might well wonder, if so, how they have been able to pursue their plans up to this point in a world in which sink capacity has been annexed by outsiders on a 'finders keepers' basis.

self-determining – struggle to make ends meet through no fault of their own. If we believe that self-determination is universally valuable, we might instead prefer *on that basis* a regime under which states with 'excess' resources remitted some of them to states with very little.

Not all defences of self-determination, though, appear to demand such untrammelled resource rights. In fact the best recent defences of self-determination – particularly functionalist or 'legitimacy-based' accounts – appear compatible with considerably less. On the most prominent functionalist view, states are only legitimate insofar as they meet their citizens' basic rights, and they ought to be able to avail themselves of the resources contained within their territories *when* those resources are necessary to meeting those rights.[12] Blomfield's argument is compatible with this more limited functionalist view: '*some* control over resources', she tells us, 'appears to be an enabling condition of collective self-determination'.[13] But endorsing this more nuanced view makes it difficult to say anything definitive about the degree of resource control we should favour across the board. Self-determination will, to be sure, be thwarted when other communities intervene to remove, destroy or restrict the use of particular resources such that a state is unable to meet the basic rights of its citizens. But we do not get from this argument any automatic presumption of control over a particular resource. We would need to make the case contextually, by providing reasons for believing that control over *this* resource was *in fact* necessary for meeting basic rights. Most likely we are pushed towards a regime where states retain control over a (fairly minimal) set of resources insofar as these prove necessary to meeting basic rights, but where the remaining 'excess' – which in many cases would be a very large excess – will be untouched by that claim, and will be vulnerable to redistributive claims on the part of communities struggling to exercise self-determination with the resources currently available to them.

On this account, then, what actually needs to be demonstrated is that *not* granting forested states ownership of the sink capacity of their forests (and not granting them, therefore, extra emissions rights) will in fact leave their ability to exercise self-determination compromised. It seems to me that that case is going to be a hard one to make. We would need to show that without these extra emissions rights, at least some of the relevant countries would be unable to meet the basic rights of their citizens. The difficulty here will be that forest

---

[12] For an example of this type of argument as applied to rights over land, see A. Stilz, 'Nations, states and territory', *Ethics*, 121/3 (2011), 575–601. Stilz remains neutral on whether the argument gives us reason to endorse control over natural resources too. But for a similar argument with regards to natural resources, see C. Nine, *Global Justice and Territory* (Oxford University Press, 2012) chs. 4 and 6.

[13] Blomfield, 'Global common resources', p.296 (emphasis added).

states with endemic poverty – or which have demonstrated an ongoing failure to meet basic rights – are also likely to be low-emitters. Countries such as Ecuador or Cameroon, for example, are currently emitting considerably below what even an equal per capita principle would grant them, and hence would already have license to ratchet emissions up considerably. Any argument for the necessity of extra emissions rights on their part therefore threatens to be superfluous. Brazil might be a different kind of case. It is already emitting close to, or beyond, what an equal per capita principle would allow, and Brazilians might feel that they should be able to emit considerably more given the sink capacity contained in their territory. But while it might be able to marshal improvement- or protection-based arguments for special rights (see below), it is hard to see how respect for its self-determination exerts any real pressure in this direction.[14]

Improvement

This leaves us with a third argument, from improvement. Agents sometimes act on natural resources in such a way as to make them more economically valuable. Perhaps they refine or purify them, or prune and tend them, in such a way as to make the benefits they supply us more bountiful or more precious. Agents can also increase the quantity of natural resources, as when they breed animals, or sow plants, or extend the range of forests. When these kinds of actions occur it seems plausible to suggest that the agents responsible for them thereby generate some entitlements. Given that the value of the

---

[14] One further argument might be invoked by the defender of self-determination, but it is again doubtful that it pushes us any closer to the conclusion of extra emissions rights. Even if exclusive rights to harness the benefits flowing from the natural resources within a state's territory prove unnecessary for meeting basic rights, wouldn't the self-determination of that state be trampled over if *other* communities were able to enter that state's territory and consume its resources? Could we really countenance agents of some UN Resource Agency dropping out of the sky, entering the territories of states and appropriating resources for use elsewhere? If not, then a state might still have the right to *block* use of 'its' resources by others even if its right to use them for its own exclusive benefit was in question. There are three things to note in response to this apparently dystopian scenario. The first is that it assumes what is at stake in many debates about territorial rights – it leans heavily, that is, on our prior acceptance of rights to control borders and to exercise exclusive jurisdiction over land. Those unpersuaded of the robustness of such rights will be less troubled by the scenario. Second, advocates of global justice might point out that there are many ways of sharing the benefits arising from natural resources – for example, by taxing their extraction or sale – which do not require us to actually expropriate those resources. But third and most important for my argument, the scenario does not apply to the case at hand. The sequestering capacity of terrestrial sinks is already consumed by outsiders, in ways that do not require them to enter the territory of forested states. Indeed outsiders could not be prevented from consuming sink capacity (which is why sink capacity counts as a collective good in the first place). The spectre of territorial incursions adds nothing to the case for extra emissions rights.

resources in question has been augmented the entitlement, on one view, would be to the increased value itself.[15] There are two kinds of reason for holding that agents can generate such entitlements through their improving actions. A direct reason would suggest that other things being equal justice is served better when agents retain benefits for whose existence they are responsible. Blomfield expresses this kind of argument in the language of desert, but we could, alternatively, describe the idea as a way of catering to the value of responsibility.[16] We could also endorse such an idea for instrumental reasons. Perhaps it is socially desirable that agents have incentives to improve resources, and to create value which would not exist otherwise. Perhaps this allows us to serve ends of justice such as the funding of public goods, or wealth redistribution. If allowing agents to retain some of the value or benefits they create when they improve resources is the best – or perhaps the only – way of doing so, we have a reason for allowing entitlements to track such acts.[17] In principle we could accept one of these reasons while remaining sceptical of the other. Often, though, both sets of reasons figure in accounts of entitlements.[18]

Such an argument could reasonably be applied to terrestrial sinks. The capacities of carbon sinks can be enhanced, as when new trees are planted which sequester greenhouse gases more effectively than older trees. Forests can also be extended by new planting, again so as to sequester more gases. When agents act in such ways it is far from unreasonable to suggest that they generate entitlements over the extra sink capacity thereby created. Communities, on this view, would not have greater emissions rights simply because they *have* forests within their territories, but because in some way they are *responsible* for the

---

[15] See e.g. D. Miller, *National Responsibility and Global Justice* (Oxford University Press, 2007), p. 218. Hillel Steiner's account also assumes a distinction between improved and unimproved resources, where redistributive claims are to the unimproved value of resources only and agents have a right to retain the value of improvements. H. Steiner, *An Essay on Rights* (Oxford: Blackwell, 1994).

[16] Blomfield, 'Global common resources', pp. 296–7.

[17] We could imagine another instrumental argument which was not conditional on improvement taking place. On this rather Humean view, one reason for according states an interest in the natural resources in their territories is the fact that those states will thereby derive an incentive to conserve those resources, or to use them most efficiently (thereby avoiding a tragedy of the commons). John Rawls briefly suggests such a justification of state ownership of natural resources in J. Rawls *The Law of Peoples* (Cambridge, MA: Harvard University Press, 1999). I consider but reject this justification of state ownership in C. Armstrong, 'Against permanent sovereignty over natural resources'. In short, it suggests giving states an interest in domestic resources, but does not rule out also granting outsiders a stake.

[18] Famously, John Locke's account of justice in acquisition both suggested that agents generate entitlements in the act of appropriating natural resources, *and* that we should accept a regime of private property rights because such a regime will generate substantially greater benefits than leaving resources in the commons. J. Locke, *Two Treatises of Government*, P. Laslett (ed.), (Cambridge University Press, 1960).

existence of at least some of the trees in those forests. A system which tailored emissions entitlements to actions taken to extend forests, or enhance their sequestering capacities, would also be instrumentally attractive. It would generate *incentives* for communities to extend forests, whereas at present all too often incentives appear to be aligned in favour of deforestation. (Notably, such an account may also have wider implications. One of the major strands of development in geo-engineering – collectively known as carbon dioxide removal (CDR) – involves agents developing technologies such as bioenergy, ocean fertilisation or enhanced weathering to remove greenhouse gases from the atmosphere. If such technologies succeed, in effect, in creating new carbon sinks, we will need moral guidance on what entitlements their schemes thereby generate. The improvement-based account would instruct us that their developers owned their sink capacity, and that under a scheme for emissions trading, for instance, they might have the right to sell that capacity on open markets).

The argument from improvement appears capable of delivering quite plausible special claims over natural resources, then. In cases where a forest has been extended from five hectares to twenty hectares through careful planting and tending, we have at least a strong *pro tanto* reason to consider the agent the owner of the benefits (including greenhouse gas sequestration) delivered by the fifteen hectares of new forest. The pressing issue, to my mind, is not with the structure of the argument from improvement (although it may possess its own problems) but instead, once more, with whether it is at all likely to deliver on the desired conclusion that states (or other communities) should be seen as the owners of the sink capacity of forests within their borders. To put it bluntly it is not, by and large, the case that the rainforests in contemporary states were planted by the inhabitants of those states. Their inhabitants have done relatively little to increase that capacity over the years. To the contrary, year on year that capacity has been steadily – and sometimes rapidly – eroded. This triggers a difficult question about baselines. If the baseline for granting extra emissions rights is to be an historical one (so that the sink capacity which belongs to forested states is construed as present capacity minus capacity at some point in the past), then there will likely *be* no extra entitlements. There might even be debts. Perhaps we could pursue some other counterfactual, to the effect that although sink capacity is now no more – or perhaps considerably less – than it once was, it is nevertheless greater than it *would have been had deforestation continued unchecked*. But that would be a quite different argument, and would not lean on improvement at all, but rather on sacrifices made in the interests of forest preservation. I consider the implications of such an argument in the final section below. In sum, however attractive the improvement-based argument looks as a forward-looking theory of the acquisition of entitlements, it appears unable to deliver a backward-looking justification of extra emissions rights for forest states.

Sharing the costs of protecting carbon sinks

If we are seeking a justification for extra emissions rights on the part of communities whose territories contain important carbon sinks, the arguments conventionally marshalled in favour of state sovereignty over natural resources turn out to assist very little. As a defence of extra emissions rights the argument from attachment is misdirected; the argument from self-determination also appears to connect rather poorly to the claim to extra emissions; and while the argument from improvement provides a plausible route towards extra entitlements, in practice it cannot take us very far in that direction. Local communities or states simply have not done enough to bring sink capacity about for the argument from improvement to be of much interest.

Should we therefore reject the conclusion that states with terrestrial sinks might be entitled to more than those without? I suspect not. What is significant, I suggest, is not whether states or other communities create or improve sinks (which they rarely do), or indeed whether they become attached to them. One thing which states with sinks such as rainforests often *do* is to incur sacrifices. Let's assume for now (normatively this is a very big assumption, though the status quo operates along just these lines) that each state has the right to use domestic natural resources in the interests of its own economic development. Assume also that while other states have a free rein to pursue economic development, states with terrestrial sinks stand to lose out on economic development opportunities if they protect or maintain those sinks. If at the same time we know that terrestrial sinks are an important bulwark against climate change, then we have a quandary. We may believe that forested states *ought* to preserve their forests (in light of the likely consequences of deforestation, we may even believe they have very firm duties of justice to do so). But we may also believe that it is unfair that those communities are required to sacrifice (some of) their development opportunities whereas others do not face similar restrictions. Selecting agents to hold duties on the basis of their capacity to perform them can make moral sense; but it can also produce distributive unfairness if those agents are at the same time required to bear all of the costs of meeting them. The problem may be intensified if forested states happen to be relatively poor, but there is an issue of justice whether or not this is so. When sacrifices are made – when forested states incur opportunity costs when they refrain from cutting down their forests to use the wood for timber, or the cleared ground for agriculture, for example – we may believe that the costs thereby incurred should be shared with outsiders.

This would be a quite different argument to any of those considered earlier. Note that the arguments grounding special claims over natural resources considered in the last section apply readily to resources whose benefits are best conceived as private goods. If I improve the value of something (as when I refine a unit of crude oil), and if I can feasibly exclude others from enjoying those benefits, then a principle which dictates that I should be able to retain those benefits can get some traction. It may make sense to treat me as the owner of those benefits (if not, necessarily, the resource as a whole). It is less obvious that such a principle will be of much use in cases where others *cannot* be excluded from consuming benefits which we are responsible for the existence of – where benefits, that is, operate as public (including collective) goods. Here, I suspect, we want to turn to a quite different argument. The principle of fairness, when applied to the provision of public goods, suggests that those who benefit from public goods ought to share in the costs of their provision (provided the benefits are worth the costs, and the costs are fairly shared).[19] The ground is then open for us to try to show how *protecting* terrestrial sinks – enabling greater benefits to exist than would occur if they went unprotected – is morally akin to *producing* public goods.

The details and precise implications of such an account would then need to be worked out. We would want to know just which costs should be shared, and when. Here reimbursing states for lost opportunity costs looks likely to be especially controversial. For an economist, the opportunity cost of rainforest protection lies in the revenues accruing to the most profitable use to which forests, or the land they occupy, could have been put. But perhaps some opportunities provide inappropriate benchmarks for compensation. Imagine, for instance, that a state would have allowed a forest to be razed for cocaine production. Would other states then have a duty to make up *that* shortfall? That seems unpalatable. More plausibly, we might compensate for the opportunity costs of activities which would generally have been permissible, but where some agents fall liable to a selective prohibition. Imagine that we are generally at liberty to pursue paid employment. But some individuals are selected, because of their strength, to protect our community from threat instead. It may make sense for them to bear this duty, but not for them to bear all of the (here, opportunity) costs of performing it. If so, the pooling of that opportunity cost is appropriate. The case for pooling the opportunity costs of rainforest protection operates according to the same logic: a general right to make the most of the natural resources within a territory is specifically truncated in the case of forest states; to pool those opportunity costs is to disentangle the duty to protect from the duty to bear the costs of protection alone.

The argument from the principle of fairness appears to enjoy good prospects of success.[20] But two significant features of the argument require attention.

---

[19] See e.g. R. Arneson, 'The principle of fairness and the free-rider problem', *Ethics*, 92/4 (1982), 616–33; G. Cullity, 'Moral free riding', *Philosophy and Public Affairs* 24/1 (1995), 3–34. For a sceptical discussion of the principle, particularly as a ground for political obligation, see A. J. Simmons, *Moral Principles and Political Obligations* (Princeton University Press, 1979), ch. 5.

[20] I unpack and defend an argument from the principle of fairness elsewhere, and deal with just these issues. See C. Armstrong, 'Fairness, free-riding and rainforest protection', *Political Theory*, (2015), online Early.

First, note that the argument, far from *justifying* territorial rights, will not apply unless we presume something like territorial rights in the first place. We need to presume, that is, that states with sinks have the right, other things being equal, to develop those sinks for it to make sense for us to share with them the opportunity costs of refraining from such development. Someone who is resolutely opposed to a state's right to use 'its' natural resources to fuel national development will require persuading on that point. Perhaps, we might say, even if permanent sovereignty should be rejected as a general principle, there are grounds for endorsing it where and insofar as it helps alleviate serious poverty, or reduce global inequalities. Some such argument has appealed to scholars of global justice in the past, and perhaps it can be reprised.[21]

Second, the *target* of the argument is quite different to the one we started with. The argument targets transnational payments for protection, treating sink capacity as a public good whose provision properly needs common funding. Its objectives coincide, roughly, with what are known as 'payments for environmental services', currently funded (in fact, chronically under-funded) through mechanisms such as REDD+.[22] However, no particular view on who ought to own, or have the right to use, the sink capacity thereby secured appears to flow directly from the account I am suggesting here. It is not obvious, that is, that the argument leads us to endorse local ownership over the emissions capacity which has been protected. The reverse might just as well be the case: perhaps the protected capacity should be treated as a *common* asset, since all of us have shared in the costs of its protection.[23] The argument for the pooling of protection costs offers no answer to that question. The argument from the principle of fairness, in short, grounds some entitlements (entitlements that

---

[21] Charles Beitz, for instance, argued very briefly that while national control over natural resources should be rejected as a general principle, there are reasons for endorsing it in the case of the poorest countries. See C. Beitz, *Political Theory and International Relations* (Princeton University Press, 1979), p. 142, n. 31. More recently Doris Schroeder and Thomas Pogge have defended state rights over biological resources on a similar basis in D. Schroeder and T. Pogge, 'Justice and the convention on biological diversity', *Ethics and International Affairs*, 23/3 (2009), 267–80.

[22] A UN initiative, REDD stands for 'reducing emissions from deforestation and degradation'. REDD+ is a consolidated initiative which also includes, inter alia, a focus on enhancing and not just protecting forests. For a discussion, see P. Kanowski, C. McDermott and B. Cashore, 'Implementing REDD+: lessons from analysis of forest governance', *Environmental Science and Policy*, 14/2 (2011), 111–17.

[23] In fact, a principle which grants ownership rights over benefits to those who have protected them from threats may produce perverse and unpalatable implications. Consider someone grinding out a match which, if left to burn, would destroy a forest. Does the match-grinder therefore derive rights over all of the benefits which the forest secures?

costs should be shared, and not borne alone). But it does not ground a wider set of entitlements over protected sink capacity.[24]

Note, finally, that if we pursued such an argument we would also need to take a position on just whom any payments for rainforest protection should percolate down to. We might simply treat states as agents and allow them to spend receipts from any protection schemes as they see fit. Or we might place constraints which sought to ensure that all citizens benefited – or that the poorest benefited the most. We might, in many real-world cases, want to ensure that the livelihoods of indigenous forest inhabitants were protected from disruption. In reality the effects of schemes to fund rainforest protection on indigenous groups have tended to be neutral at best, in the sense that they have not seen the benefits which have often been promised. In many cases they have suffered profoundly negative effects, including being excluded from the forests in which they have long lived in case their customary activities affect the way in which a forest's sequestering capacity is estimated.[25] Given this tension, we would need to take a view on the difficult question of whether payments for protection should be calibrated to serve local justice goals (such as equalising incomes, or protecting the poor or the local indigenous populations), or whether they should be calibrated by *results* in protecting sink capacity. It would be tempting, but too easy, to assume that such diverse goals as protecting sink capacity, maintaining biodiversity and protecting the ways of life of indigenous peoples all point straightforwardly towards the same institutional solutions. In practice a focus on maximising sink capacity has often led to backward steps in terms of biodiversity and indigenous rights. Developing policies which might pay proper heed to a more diverse set of goals – is therefore a real challenge.

---

[24] Of course, it might be the case that we could approach the latter conclusion in practice. If states are to be paid to protect terrestrial sinks within their territories, then under a quota-trading system they could of course use the money gained to buy themselves extra emissions entitlements. It seems to me to be difficult, in advance, to estimate just how much in the way of greater emissions entitlements states with terrestrial sinks would then end up with. We can affirm that the capacity which might be bought would not exceed that protected (a fact determined by the proviso that the benefits of protection must be worth their costs). But there is no reason to expect them to match the capacity of sinks themselves; they might fall well short of such levels.

[25] See e.g. J. Phelps, E. Webb and A. Agrawal, 'Does REDD+ threaten to recentralize forest governance?', *Science*, 328/5976 (2010), 312–31; A. Larson, 'Forest tenure reform in the age of climate change: lessons for REDD+', *Global Environmental Change*, 21/2 (2011), 540–49.

4 Exporting harm

Jeremy Moss

In this chapter I will discuss an aspect of the problem of how to divide the moral and budgetary responsibility for greenhouse gas (GHG) emissions. By budgetary I mean the world's remaining 'carbon budget', the amount of carbon dioxide-equivalent (CO_2-e) that can be emitted if we are to avoid dangerous climate change. I want to establish that there is a prima facie case for allocating responsibility for the harms caused by 'exported emissions', as well as those that are produced within a country's borders. This is not a complete determination of the carbon budget problem by any means, but a step towards its development. This chapter sets out some of the factors that determine a country's carbon budget and its responsibilities for harms caused as a result of emissions, and argues that the current methods for allocating emissions and responsibilities for their harms are inadequate and more complex than they appear.[1]

Without immediate action to reduce GHGs it is highly likely that the world will experience dangerous climate change in the near future. The consequences of the global mean temperature rising by or beyond 2°C have been extensively discussed and I will not repeat them here. In order to avoid dangerous climate change we need to limit the amount of CO_2-e that is released into the atmosphere to keep temperature rises to less than 2°C.[2] Given that we have already released a large amount of CO_2-e we have a very small carbon budget with which to operate. Following the scientific literature, I will assume that the period in which we have to 'spend' the remaining carbon budget is between now and 2050.[3] According to some estimates, to have a 67 per cent chance of avoiding a 2°C temperature rise

---

[1] I would like to thank the audiences at the University of Melbourne, the 'Who is Responsible for Climate Change' workshop (ANU, CAPPE) and at the Australasian Association of Philosophy conference for their feedback on this chapter. The research has also been supported through an Australian Research Council Future Fellowship on 'Climate Justice'.

[2] Intergovernmental Panel on Climate Change (IPCC), *Climate Change 2013: the Physical Science Basis. Contribution of Working Group I to the Fifth Assessment Report of the Intergovernmental Panel on Climate Change*. T. Stocker *et al.* (eds.) (Cambridge, UK and New York, USA: Cambridge University Press, 2013).

[3] IPCC, *Climate change 2007: Impacts, Adaptation and Vulnerability. Contribution of Working Group II to the Fourth Assessment Report of the Intergovernmental Panel on Climate Change*, M. L. Parry *et al.* (eds.), (Cambridge, UK and New York, USA: Cambridge University Press, 2007); German Advisory Council on Global Change (WBGU), *Solving the Climate Dilemma:*

the entire global emissions from fossil fuel sources cannot exceed 750Gt for the period 2010–50.[4] After 2050 only a very small amount of CO_2-e can be emitted. If the probability of staying within the 2°C guardrail were increased to 75 per cent, the budget would reduce to 600Gt. Moreover, in order to avoid having to reduce emissions at an unrealistic rate, reductions must begin immediately with peaks between the period of 2011–20.[5] For instance, a peak in 2011 would have resulted in an annual reduction rate of 3.7 per cent, a peak in 2015 of 5.3 per cent and 2020 of 9.0 per cent.[6] Raising the probability of avoiding the temperature rise would obviously increase these targets.

As these figures show, the situation is bad now and getting worse with every year in which no overall agreement – let alone effective action – to reduce emissions is reached. Moreover, many countries are 'spending' their likely carbon budgets and may well be in deficit in the near future. While we will discuss some of the ways in which we can divide the global carbon budget below, for purposes of illustration, assume that we allocate emissions on a per capita basis with 2010 as the reference year for population data. On this scenario, we would divide the carbon budget by the world's population, allocate emissions to countries on the basis of their population and estimate an annual emissions rate on the basis of 2010 population levels.

On this formula, the USA would be allocated .85Gt per year whereas it currently emits 6.1Gt. It would spend its budget in six years. China, on the other hand, would have twenty-four years, which is obviously better but still well above the emission rates that are required. If we assume a global allocation of individual emission budgets, each person would be allocated around 2.7 tonnes (t) of CO_2 per annum (pa) for forty years.[7] This is around 25 per cent of what New Zealanders spend every year (8.6t pa), and the UK (9.4), which is better than Australians (17.3t pa) and much better than the world's leading emitters, the Qataries at (60t pa).[8] While there are of course many other ways to distribute emissions, this example illustrates both the magnitude of the task of reducing emissions and the importance of getting the distribution right according to a defensible criterion.[9]

In order to allocate emissions between countries we need to know which emissions a country is responsible for. The current Intergovernmental Panel on

---

the Budget Approach (Berlin, 2009), available at www.wbgu.de/en/publications/special-reports/special-report-2009, last accessed 10 October 2014.

[4] WBGU, *Solving the Climate Dilemma*, p. 28. This figure excludes emissions from land use.

[5] Annual anthropogenic GHG emissions have been running at around 49Gt. So the annual CO_2 budget is likely to have decreased markedly since the WBGU report. N. Höhne, J. Kejun, J. Rogelj, L. Segafredo, R. S. da Motta and P. R Shukla, *The Emissions Gap Report 2012: a UNEP Synthesis Report*. November 2012.

[6] WBGU, *Solving the Climate Dilemma*, p. 16. [7] WBGU, *Solving the Climate Dilemma*, p. 28.

[8] K. A. Baumert, T. Herzog and J. Pershing, 'Navigating the numbers: greenhouse gas data and international climate policy', *World Resources Institute* (2005), p. 22.

[9] See S. Caney, 'Just emissions', *Philosophy and Public Affairs*, 40/4 (2013), 255–300.

Climate Change (IPCC) guidelines that are used by the United Nations Framework Convention on Climate Change (UNFCCC) state that national inventories of GHG emissions 'include greenhouse gas emissions and removals taking place within national territory and offshore areas over which the country has jurisdiction'.[10] What they call 'scope 1+2' emissions are those emissions that are produced within a country's borders by various types of activity (industrial activity, transport and so on). Scope 3 emissions are the emissions that are produced by the commodities when they are generated outside the country's territory. These latter are not part of the country's emissions budget. To give an example, if a country exports coal the emissions that are generated in extracting the coal and transporting it to a port are part of that country's emissions budget because they occur within its territorial boundary. The emissions that are produced when the coal is burnt are part of the actual budget of the country that burns them.

Accounting for emissions in this way seems an intuitively plausible way of dividing the emissions budget. If we think of this question in terms of the scope 1–3 framework, the moral and budgetary responsibility seems to fit nicely into the framework set down by the IPCC. Countries have budgetary responsibility for scope 1 and 2 emissions but not for scope 3 emissions. Similarly, any harms caused by scope 1 and 2 emissions are that country's responsibility. What happens as a result of what others do with exported fossil fuels, for instance, is a matter for the recipient country.

However, there is a plausible case for claiming that countries have prima facie responsibility for at least part, and sometimes all, of the harms caused by their scope 3 emissions, which might entail: including them in their carbon budget, curtailing their exports, or being liable for compensation for the harms that have been caused. Consider the following analogies. Suppose that a country produced and exported large amounts of tobacco to a developing country that did not have health warnings for smoking. Given what we know about the links between smoking and death and disease, the exporting country is plausibly implicated in the harm caused and morally responsible for at least some of that harm. Another example concerns hazardous waste. Where one country knowingly ships its dangerous medical or industrial waste to a country that has low or no standards for its safe disposal, we can say that the exporter bears some responsibility for harms that may result when the waste is not

---

[10] 'The *2006 IPCC Guidelines for National Greenhouse Gas Inventories* (*2006 Guidelines*) were produced at the invitation of the United Nations Framework Convention on Climate Change (UNFCCC) to update the *Revised 1996 Guidelines* and associated *good practice guidance* which provide internationally agreed methodologies intended for use by countries to estimate greenhouse gas inventories to report to the UNFCCC ... National inventories include greenhouse gas emissions and removals taking place within national territory and offshore areas over which the country has jurisdiction.' (p. 14). H. S. Eggleston, L. Buendra, K. Miwa, T. Ngara and K. Tanabe (eds.), *IPCC Guidelines for National Greenhouse Gas Inventories* (Japan: IGES, 2006), Sct 1.4: 4.

properly disposed of. This is still likely to be the case even where there was consent from the importer to take the waste. More obvious still is the case of uranium exports. There are good reasons why many countries place restrictions on the end destinations of their product. The risks of weapons proliferation, accidents at reactors, storage issues and so on are just too great with some countries to countenance an export programme. Should one country knowingly export uranium to another country where safety is lax we could rightly accuse it of being irresponsible and having a share in the blame if an accident were to happen. We could make similar analogies with the exports of other products, such as live animals or dangerous industries that use asbestos.

What all of the cases above have in common is that they cause harm in a morally significant and blameworthy way. For the purposes of this chapter I will assume that causing harm should provide a powerful and important constraint on our actions. It gives us a prima facie reason to restrain our actions and/or to be liable for the consequences.[11] This is not to say that there are not all things considered reasons why harming someone might be permissible in certain circumstances, but that these reasons will have to be argued for. A harmful act may be perpetrated in a situation where the agent simply has no choice but to harm someone. An agent may also harm someone unintentionally or by dint of unforeseen circumstances. These and other reasons may mitigate the responsibility for the harm and its consequences. I will explore some of these reasons in relation to the export of fossil fuels in what follows. Specifically, I will discuss whether some of the all things considered reasons such as that the overall benefit of emitting practices ought to count; claims that harms that involved the consent of both parties in a transaction cancel responsibility; as well as claims concerning intention, broken causal chains and having no alternative.

To return to the cases above, fossil fuels are a commodity which we know causes harm when we export it for its standard uses. We know that gas or coal exported from Australia or Brazil to China will be used to generate energy and that this process will release GHGs into the atmosphere and the links between increasing GHG in the atmosphere and harms caused is strong. Moreover, the harms that this chain of events potentially causes to people everywhere impact on their significant interests, these include: increased vulnerability to disease, crop failure, water shortages and the impacts caused by severe weather events. Even though the harms of climate change will fall on all

---

[11] One of the plausible ways in which we can understand the badness of these practices is by invoking the harm principle. There are numerous different accounts of the harm principle that might fit this context. For other similar discussions see: H. Shue, 'Exporting hazards', *Ethics*, 91/4 (1981), 579–606; E. Cripps, 'Climate change, collective harm and legitimate coercion', *Critical Review of International Social and Political Philosophy*, 14/2 (2011), 171–93; J. Feinberg, *Harm to Others: the Moral Limits of the Criminal Law* (Oxford University Press, 1984).

societies, the harms will likely fall on those who are already significantly disadvantaged to a greater degree. The harm caused is like the harms caused by tobacco in morally relevant respects in that exporting countries contribute to knowingly harming the significant interests of others in ways that could, in many cases, be avoided. In wealthy countries such as Norway and Australia, the export of fossil fuels is not the only means for those countries to maintain a high standard of living. They may be significant to the economy, but they are not the only contributions to what is already a high standard of living. If it is the case that exported fossil fuels are exported in the knowledge that they cause significant harm to significant human interests and the practice could be avoided, then resource-exporting nations have a prima facie responsibility for the harms that they cause through the export of resources. I will address what choices exporting countries really have and what their actions should entail below. But for now we should note that accepting responsibility might entail compensation, a commitment to cease contributing to the harm where necessary and the possibility of including at least some of the scope 3 emissions in the exporting country's carbon budget.[12]

Some will argue that in relation to many of the problems discussed here, we should draw on positive duties as reasons for restricting or changing our behaviour in relation to climate change and in assisting others to do so. I acknowledge the force of these claims and do not rule out such arguments. But I will focus on the claims arising from negative duties here because they seem the best able to capture the situation regarding responsibility for a significant set of emissions and the carbon budget and because of their moral stringency. The claims that arise from violating people's significant interests are a particularly stringent kind of claim because of the weight we associate with this kind of harm. The moral stringency of negative accounts of duties does not mean that there is no case from positive duties. But this argument is independent of such accounts.

Before looking at responses to this basic argument, I would like to note two closely related cases of emissions counting. A parallel case to the one of fossil fuel exports concerns where countries export their emission producing industries. A country might have decided to move its carbon intensive factories it uses to produce its goods to another country because it is cheaper and easier to produce goods there, and because it transfers the carbon burden to that other country. One consequence of doing this is that it also transfers the emissions that are produced in the manufacture of goods to the manufacturing country's carbon budget. I cannot fully develop the implications of this case here, but a

---

[12] I leave aside the question of agency. I am not interested here in the kind of questions surrounding the best understanding of agents. I will speak mainly of countries just to get the discussion of the problem underway. I also leave aside the issue of whether there are entitlements to coercively enforce a cessation to the harm.

number of points seem to follow from what I have argued above. The first is that where harm is caused similar sorts of considerations to the ones relevant to fossil fuel exports apply. If there is harm caused by these emissions and where the harm is avoidable and inflicted, then the outsourcing country has a reason to assume some of the responsibility for the harms and possibly for the emissions. There is no significant moral difference between outsourcing activities rather than commodities. As I will discuss below, there might of course be many reasons why the responsibility of the exporting countries might be diminished. The countries where the factories are now located may want those factories because they provide jobs. The new location countries may be happy to have the extra carbon in their budgets (they may also have room for it), so the harms may be mitigated. But where it is a case of their being in a desperate situation because of poverty and underdevelopment there might also be some infliction of harm just as there is in the cases above. If this is the case, then the exporting country may not be absolved of all responsibility.

An interesting version of this case occurs where one country does not relocate its factories but nonetheless sources many of its goods from another country via contracting with them for the production. For instance, in a recent study Bin Shui and Robert Harris estimate that as many as 14 per cent of China's GHG emissions between 1997 and 2003 were the result of manufacturing goods for export to US consumers. They also estimate that aggregate US emissions would have increased by 3 to 6 per cent if those goods imported from China had been produced in the US.[13] Here there might also be responsibility entailed by the purchase of goods with embodied emissions. However, the responsibility in theory as well as practice is much harder to determine. Attributions of responsibility will depend on the role that each country played in causing the goods and their associated GHGs to be produced. I will not go into all the issues here such as whether the supplying of a demand or the creation of a demand is the most causally relevant factor, or whether there is often little choice for developing countries but to produce goods for richer countries if they are to increase their level of well-being.

All things considered: is it really a harm?

Granting that we ought to avoid causing harm, we might question whether the harm in this context really is like the more clear-cut cases of exporting medical waste or uranium. I'd like to start with some all things considered objections. These are obviously not objections to not harming per se, but to how it is

---

[13] B. Shui and R. Harris, 'The role of CO_2 embodiment in US–China trade', *Energy Policy*, 34/18 (2006) 4063–8.

applied in this case. There are two types of objections that I wish to address: respectively, 'isolation' and 'offset' arguments.

To start with the first type of all things considered judgement, someone might object along the following lines to my proposal. They might argue that while there is a clear climate harm caused by exports, overall these harms are balanced out by the benefits that generating, say, power from coal brings. Coal-fired power plants have allowed developing countries in particular to produce cheap reliable electricity that has driven their economic growth and helped lift millions out of poverty. Evaluations of emissions levels should not be made in isolation from these broader considerations of justice: such decisions must be taken all things considered (the isolation argument).[14] Yet if this is true, it ignores something important about climate harms: the structure of benefits and harms is not isomorphic. In similar standard cases often invoked in connection with a harm, such as workers who knowingly accept dangerous jobs for high pay, while there is certainly a harm or a high risk of harm involved for the person performing the job, the benefits might outweigh the risks because there is no other work, they have a family to support and so on. But the harm of climate change is not like this. Exporting fossil fuels certainly does produce benefits and harms in the county where the fuel is consumed. Using coal to generate electricity helps those who consume the electricity by allowing them to heat their homes or get access to jobs. However, the harm of climate change potentially falls on everyone on Earth to varying degrees because the effects of a change in the climate are diffused throughout the world. Producing emissions in one place does not just alter the climate in that place alone, although it may do so, but elsewhere on the planet as well. So the kind of response that claims that the benefits outweigh the harms ignores the non-isomorphic nature of the benefits and burdens in the case of climate change. There could, of course, be some trickle down benefits from increased global trade or technological development, but by their nature these benefits are only a small fraction of the benefits created and probably not likely to compensate for increased exposure to extreme weather events, sea level rise, reduced rainfall or bushfires.

A second all things considered claim is that we really should not be looking at emissions from exports in isolation from a country's overall carbon footprint. What we are really interested in is not the raw amount of emissions, but whether a country offsets its emissions in some way by maintaining carbon sinks either domestically or abroad, buys emissions permits or has a large population that might explain its level of emissions (offset argument). The claim is that what matters is not the total amount of emissions, but the amount

---

[14] For arguments of this type see Caney, 'Just emissions'; S. Scheffler, *Boundaries and Allegiances; the Problems of Justice and Responsibility in Liberal Thought* (Oxford University Press, 2001), p. 166.

of overall emissions after offsetting or the purchase of permits are taken into account. Imagine that in addition to being an exporter of fossil fuels a country was also a creator of carbon sinks that absorbed at least as much carbon as was emitted using my version of the carbon budget. In this case, what is relevant is not the amount emitted but a country's overall carbon budget. This is also true if we factor in trade in emissions permits. In taking account of this objection, we should still note that while the general point is correct, there is still a prima facie case for countries to count at least part of their exported carbon and accept proportional responsibility for the harms it causes. The fact that those harms may be cancelled out by other factors that are part of an all things considered judgement does not alter the prima facie claim.

Regardless of some of the all things considered factors, it remains true that, if we are to acknowledge the claim, we ought not to harm others deliberately in significant ways in order to further our own ends. To go back to our earlier examples, if this were not the case, then it may not be an injustice to export uranium in a profitable way even though we knew that it would end up in a leaky reactor or in the hands of illegal weapons manufacturers. Nor would it be unacceptable for wealthy countries to export medical waste to countries where it would end up in unsafe storage. All things considered benefits are not enough to annul the prima facie injunction against these kind of practices, where the party responsible for the harm could avoid doing so. While it might be the case that the kind of all things considered judgements mean that, on balance, countries can stay within their carbon budget even if we count emissions from exports. But my point is that there is still a prima facie reason to include them in our national moral and carbon equations in contrast to our current practices. This means that we have a reason to take responsibility for emissions from exports and the consequences of doing so are likely to be significant, as the figures for the rate of adjustment cited earlier attest.

A different type of objection to using harm in this way is that it is simply 'mere causation', that we have failed to establish 'cause-in-fact' as it is called in the law. For example, if the actions of coal exporters were merely a minor prior causal condition of the harm of climate change, then the causal link between the actions of the exporters and the harm would be minimal. When a bank robber makes a successful escape after a heist, and does so thanks to the well-constructed roads leading away from the bank, we are unlikely to hold the road planners responsible for any part of the crime. The roads were certainly essential to the success of the robbery, yet the contribution of the planners has little moral relevance. There is substantial discussion of how best to establish cause-in-fact in the legal and philosophical literature, focusing on whether the action needs to be necessary to the relevant outcome, (but-for causes), part of a set of conditions sufficient to the outcome or whether the relevant action makes a substantial contribution to the outcome. I will not repeat these arguments here.[15] I will assume that there is a substantive type of morally significant causal relation in the case I am describing as there is for the other examples of harmful exports mentioned above. I will also not directly consider cases where there is a question of whether one emits CO_2 in the context of very many other agents emitting similar amounts overdetermines the harm. Overdetermination cases of climate harms are important but I will assume that major exporters of fossil fuels are part of a causal chain that can be considered a significant contribution to climate harms.[16] What I will consider is whether the type of causal relation involved in the export case is the kind of case to which we should to attach moral responsibility that ought to result in remedies such as ceasing to export or compensation. These fall into several main types: whether the act was intentional, involved consent and was indirect. There are then the further issues concerning the degrees of or levels of responsibility and what kind of moral and political consequences ought to follow from this type of harm.

Responsibility for harm: limiting conditions

A standard defence against invoking harm as a reason not to act in a particular way is that the apparent victim consented to the harm. With the case of the worker who accepts a dangerous job, on the most charitable interpretation he accepts the job voluntarily. But even allowing a charitable interpretation and putting aside the issue of whether the worker has any feasible alternative, the analogy still does not hold with the fossil fuel export case because the harmed do not have an option of accepting the harm. As we noted above, if one country emits GHGs this adds to global warming, whose effects are diffused across the globe. Those outside the export-related transaction do not get to consent to the imposition of this particular harm. Whereas, in the case of the worker who accepts the dangerous job, they might have weighed the harm against the benefit and decided that it was worth accepting. Not only does this balancing make a difference, but the ability to be able to make judgements between alternatives puts the worker in a situation that is superior to the person or

---

[15] A. M. Honore, 'Necessary and sufficient conditions in tort law', in D. G Owen (ed.), *Philosophical Foundations of Tort Law* (Oxford: Clarendon Press, 1995), pp. 363–85; J. Stapleton, 'Legal cause: cause-in-fact and the scope of liability for consequences', *Vanderbilt Law Review*, 54 (2001), 941–1009; R. W. Wright, 'Causation in tort law', *California Law Review*, 73 (1985), 1737–828.

[16] For a discussion of these issues see: J. Nolt, 'How harmful are the average American's greenhouse gas emissions?', *Ethics, Policy and Environment*, 14/1 (2011), 3–10; W. Sinnott-Armstrong, '"It's not *my* fault": global warming and individual moral obligations', in W. Sinnott-Armstrong and R. B. Howarth (eds.), *Perspectives on Climate Change: Science, Economics, Politics, Ethics*. Advances in the Economics of Environmental Resources (Amsterdam: Elsevier, 2005) vol. V, pp. 285–307.

country that simply has the harm imposed on them. This point also applies to the all things considered argument above. Even if third parties do gain in some way from increased global trade that results from cheap electricity brought about by exports, they do not have any say about these benefits and how they might rank them in relation to other benefits.

One obvious potential limitation to liability for this kind of harm is that exporters of fossil fuels do not knowingly set out to harm people by their actions even if they are aware of the risks. Let us assume that this is indeed the case and that sellers of fossil fuels do not do so intentionally to harm others. This fact alone will not be enough to absolve them of some degree of responsibility. For a start, it is also true of those who burn fossil fuels in their power plants. They presumably do not do so in order to deliberately harm others. If this was the only criterion for attributing responsibility, then no one would be relevantly responsible for the kind of harms caused by emissions from sources such as coal burning power stations. Whereas, if exporters of coal are acting in disregard of a known and serious risk of harm resulting from the use of their fuels, they plausibly still bear some degree of responsibility for the harms. Indeed, exporters know the use to which the fuels will be put and the kind of harms that are likely to occur as a result. This is partly because a commodity such as coal has a single use. We know the uses to which coal will be put – being burnt to generate power – and that these uses are harmful to the climate. The exporter case is more like the seller of uranium case mentioned above. For example, were a country to sell uranium to a dictator who then, predictably, caused a nuclear accident, the seller should shoulder some of the blame for the harm. While the harm is indirect, nonetheless the seller of uranium not only makes a crucial contribution to the harmful situation, they do so while recklessly ignoring the risks that are involved in selling the commodity. There is an analogous situation with the exporters of fossil fuels. Without their contributing the fuels the harm would not occur and they know that harms will be caused from the intended use of the fuels. In this and the export case responsibility is incurred through consciously ignoring the likely risk of contributing to the direct cause of a harm.

Nonetheless a country might object that they have every right to extract and export fossil fuels that fall within their territory. Leaving aside a range of typical considerations concerning ownership and national sovereignty, two considerations are particularly relevant here. Let us assume that countries do have ownership rights to things that fall within their territory. What they do not have is a right to use these resources in a way that obviously harms others in the kind of serious ways that we have been considering. But there is also another powerful constraint on what a particular country, especially where that country is a major exporter, can export and that is a fair shares consideration. According to the International Energy Agency (IEA), in order to avoid dangerous climate change we can only extract and consume one-third of the total known global reserves of fossil fuels. This means that we have to be careful and systematic in how we go about utilising and monitoring those reserves that we have. But this is the opposite of what is currently happening. As it stands, any country can dig up what it likes and consume it or export it. This is a cause for concern for a number of reasons. The first and most important reason is that it takes no regard of the needs or claims of other countries to use up the stores of fossil fuels. Take a simple case of a large country with a high standard of living that has an abundance of fossil fuels and has exported its reserves of fossil fuels for sixty years. Compare it to another group of countries that are extremely poor and have just discovered gas fields that, in the absence of other sources of income, are their only way out of poverty. It seems intuitively unfair that the first country should have no limits on the fuels that it can export and the quota of the world's budget for fossil fuel emissions that it can use. Even where there is no agreement on the exact determination of what a fair share is likely to be, many countries will be over the limit of what any plausible fair share is likely to be. The issue here is not one of who should own the resources, but with who is harmed or benefitted by their use.

Some might respond that while there are issues of fair shares to be addressed, the fact that there is no legally binding agreement between countries to regulate these shares ought to mean that we cannot blame countries for allowing companies to extract and export unlimited quantities of fossil fuels.[17] Yet the fact that there is no legally binding global agreement should not preclude exporting countries accepting some responsibility for the harm to which they are likely to contribute. So fair shares arguments give wealthier countries in particular a further reason not to export their fossil fuels. Not only are they harming others by creating climate harms, but they are engaging in potentially unjust distributional arrangements.

The scenarios for accepting responsibility that I have discussed have assumed that the country in question can feasibly avoid exporting fossil fuels. As far as avoiding harm is concerned this makes a difference. Insofar as we ought not inflict harm on others, one factor that might mitigate the responsibility of the exporting country is whether they had a feasible alternative. Certainly for wealthy exporting countries, the benefits gained are on top of an already relatively high standard of living. Many exporting countries could not export their fossil fuels and still not end up without the means to survive and maintain a similar standard of living. The situation facing the USA or Brazil is not one of total collapse if certain kinds of harmful fossil fuel exports were ceased. For some countries such as Timor Leste, it may have little choice but to export fossil fuels if it is to climb out of extreme

---

[17] For a view of this kind see Feinberg, *Harm to Others*.

poverty. This kind of example is a challenge to whether the harm is avoidable in a meaningful sense. Climbing out of extreme poverty may well provide some justifiable reasons to mitigate some of the consequences of causing harm. There seems a clear moral difference at least between our assessments of countries emitting to avoid extreme poverty and those countries that emit to maintain an already high standard of living. Yet even here, the longer the world goes without a clear agreement and plan to reduce emissions even poor countries will have to lower their emissions budget if we are to avoid dangerous climate change. Moreover, necessity as a mitigating factor will not be as relevant for countries like Norway or Australia, where fossil fuel exports occur on top of an already high standard of living.

In contrast to the Timor Leste case, some may object that it seems unfair to blame the exporting countries for the bad habits of well-off importers. Take the example of where one country exports oil to another country so the latter's residents can drive their cars. They may need their cars for essential purposes but also use them for holidays, recreation and simply because it is easier than walking or taking public transport. Why should the exporting nation be blamed for the emissions that are produced by these car drivers who often use their cars simply for convenience or for non-essential purposes? Unlike some of the earlier cases, in this case it is not just the exporters who are engaging in emitting primarily for luxury behaviour but the importers as well. Cases of this kind may shift the all things considered judgements about responsibility towards the importers. Nonetheless, exporters of fossil fuels still play a role in contributing to the harms caused by emissions in cases of exporting fuels for luxury use. If this is the case, then they should also bear some of the responsibility for the harms even it is less than they might bear for exporting for essential uses.[18]

Apportioning consequences

I now want to turn to the kind of consequences that we could expect if this framework for deciding on the harm caused by exported fossil fuels is true, especially for how it ought to guide the exporting country and in determining their carbon budget. There are two types of consequences, one related to the carbon budget and the other to the general responsibility for harms. My aim here is not to provide a fully worked out conception of what a country's carbon budget would look like after factoring in the emissions produced by exports or its complete set of moral responsibilities, but to offer some assessment of the direction in which this argument takes us.

So far I have assumed that the exporting country is wholly responsible for the harms that its exports cause. Some may object that this is unlikely to be the

---

[18] I thank Jonathan Pickering for alerting me to this point.

case. In a complex trade relationship between two genuinely independent countries even if they are unequal, it is likely that the responsibility for the harms caused by GHG emissions does not rest solely with the exporting country. So too with the increases in counting carbon. Why is this? To return to one of our earlier examples, if one country exports uranium in a situation where there are known risks of it being used in a dangerous or unsafe way, then they are in part responsible, but not wholly so. The recipient country is likely to be also at fault for its dangerous practices. This is also the case with exporting fossil fuels. In reality, determining exact levels of responsibility is likely to be a complicated process.

Even though my discussion here does not provide an exact formula, it does provide some guidance about the kind of actions that a country or company ought to take. The first and most obvious consequence is the impact on an exporting country's carbon budget. If all or a proportion of the exported emissions are counted, then countries will be able to produce less emissions domestically if they are to stay within their budgets. As a result, they may have to make large adjustments to practices that cause emissions domestically, such as their stationary energy or transport sectors. As I have indicated, a country is unlikely to be wholly responsible for its emissions exports, but even a relatively small allocation to its domestic budget would require changes given that many exporting countries are already over or at their emissions limit. In some cases this will not entail much sacrifice, but in others it may be very costly. An increase in the carbon budget of one country will of course mean an equal decrease in that of another where the increase comes from counting exports.

I should note here that this point does not rule out measures such as offsetting. A country may conceivably emit a great deal but plant vast forests of carbon-absorbing trees or take abatement measures such as carbon capture and storage. I do not rule out this response as a way to achieve a diminished carbon budget. What we care about is how much CO_2-e gas is in the atmosphere. However, we need to note that any such program will have to at least observe the constraints of not harming others. Offsetting measures should not cause harm to others in unacceptable ways such as conspiring with other governments to remove people from their traditional land or by buying all the cheap emission permits when a country is at its least developed and leaving it little room for its own development associated emissions. Offsetting must also be effective and long-term. Offsetting emissions by creating forests in high bushfire risk areas or in countries subject to uncontrolled land clearing is clearly undesirable. Second, we should also note that purchasing permits or establishing offset programmes are themselves costly and so potentially of considerable impact on the countries concerned. These responses aside, where emissions from exports are counted this adds a strong motivation for a country to take measures to reduce its emissions. This is especially so where either its emissions are already high, or its counted emissions from exported fuels are high, or both.

A more significant consequence of accepting the harms that are caused by exporting fossil fuels is the duty it generates to stop causing the harms. Where the external harms that are caused by exported emissions are inflicted and affect the significant interests of others, there is a prima facie reason to stop exports where they are part of the cause of the relevant harms. If we can agree that it is unjust for people to knowingly continue practices like exporting fossil fuels where it significantly and avoidably harms the interests of others, then absenting all things considered judgements to the contrary, there is a reason to curtail those practices. This is especially the case given that the harms in question are inflicted on third parties who derive little or no benefit from the exports themselves. This constraint should at least inform the decision-making process of countries who are considering commissioning new fossil fuel export developments. Accepting that there is likely to be some exporter responsibility for the harms caused by emissions provides a reason to limit or not proceed with such developments.

Harming others also raises the issue of how to remedy harms that have already occurred. Where countries do knowingly harm the significant interests of other parties in a manner that is consistent with not harming others, then I will assume that they have a prima facie case to offer redress for those harms. Where this takes the form of compensation one of the issues that any compensation scheme must address is how to compensate for harms when the impacts of climate change are so diffuse. As I have mentioned, the burning of fossil fuels is contributing to harmful climate change, but establishing direct causal links between the operations of one power plant or even the contributions of a small country and specific harms in another country such as crop failure, damage from extreme weather events and so on is difficult. This means that we have to be careful concerning how we compensate third parties in the right way.

While the nature of the harm does raise problems, they are not insurmountable and do not pose a barrier to compensation. For instance, a country that is guilty of allowing fossil fuel exports could conceivably contribute to a global trust that was set up to address these kinds of harms. In relation to climate change, several of these kinds of mechanisms already exist. The trusts could then allocate compensation in an effective way. But note that if the aim is to compensate for harms the activities of these trusts should not be confined to fixing the climate. Their purpose could include climate-friendly projects but should primarily aim at addressing the harms caused by climate change, such as loss of arable land, housing or the harms caused by exposure to extreme weather events.

What this also points to is an issue about the kind of compensation that ought to be provided. While I cannot address this issue in any great detail here, if we are concerned with compensating people to remove the disadvantage that has been caused, if we are to respect the autonomy of those who have been harmed we should focus on giving them the means to improve their situation and not externally chosen ends. Unless there is some sort of bargain or consent to do otherwise, they ought to be compensated for the ends that have been affected. It seems insufficient to claim that there are unexpected new goods, such as increased prosperity or economic trickle down. This seems not to be respecting autonomy of those on whom the benefits and burdens fall. There is a 'legitimacy deficit' in the sense that those who receive the benefits have no say in whether they want them, would prefer other benefits or would rather not have the harm even if they get some benefit.

Respecting the ends of the people harmed raises another issue: if this kind of compensation is possible why is it not acceptable to simply continue the practice of exporting harm but with adequate compensation? The main reason is that there is still something intuitively objectionable about imposing possibly severe harms on people without their consent and then trying to compensate them. Of course, there are instances where we compensate people for a loss that is imposed on them. An obvious example is the demolition of a house to make way for an important piece of infrastructure. Even though such cases are not without controversy, they at least occur in the context of a political structure where there is recourse to a means of arguing about the compensation, potentially benefitting from the new amenity and so on. These kinds of considerations also apply to resources directed at adaptation. Some may argue that a way around the injustice caused by the harm is by assisting people who have been affected to adapt. This may well be an appropriate form of action for harms that have already been caused. But for similar kinds of reasons to the ones above, we ought not to think that the possibility of assisting with adaptation measures justifies continuing the practice of exporting harm.

One further point that mitigates against continuing the practice of fossil fuel exports is that the benefits often accrue to the well-off and the burdens to the disadvantaged. We have seen how the burdens associated with climate change will most likely hit many of the poor hardest because they are already on the edge of survival and do not have the means to combat further disadvantages. That the extra burdens associated with climate change are caused, at least in part, by practices that add or maintain the wealth and well-being of the already advantaged, is an additional reason to think that it is an injustice. As we have seen in the case of wealthy countries, because these harms *could* be stopped at relatively little cost, the reasons for not doing so are stronger and the continuing harm to the poor harder to justify.

Conclusion

What I have tried to show is that we need to modify our framework for understanding how to account for a country's carbon budget and for how we think of the moral responsibilities that countries have for the harm caused by the emissions they export. While there are a range of all things considered factors that will determine a country's responsibility for harm and their carbon budget, there is a strong prima facie duty to accept the harms that are caused by exported emissions and to cease to inflict those harms. While there might be several all things considered reasons for a country to continue exporting emissions, in reality this is unlikely to be the case for wealthy exporters of fossil fuels. This new framework for understanding a country's carbon budget and for assessing its responsibilities better allows us to account for who is responsible for emissions.

5 What's wrong with trading emission rights?

Axel Gosseries

Introduction

Climate change is one of the most serious challenges of our times.[1] How to make sure that action be taken is a key question. Whether fairness in climate policy can contribute to reaching agreement on effective measures, on top of being important for its own sake, is central as well. Other contributions in this book have addressed various dimensions of the problem. Here, I will focus on a specific one. I will look at tradability, i.e., whether emission rights typically granted to countries should be tradable. I am not claiming that tradability is the most important issue in practice. However, given the daunting nature of the challenge, each and every possible contribution to a fair and effective regime should be considered. Tradability is of course a very large question generally falling under the 'commodification' or 'marketisation' heading, i.e., whether a good or service should become or remain a commodity, exchangeable on a market. My aim will be to focus on tradability of a specific entitlement, i.e., emission quotas. Also, as I will have limited space, I will not go into the history of the idea of tradable quotas and the diversity of its implementation.[2] I will also not aim at comparing the virtues of tradable quotas schemes with the ones of the other main 'incentive-based' regulatory alternative, i.e., Pigovian taxation. Nor will I aim at a systematic comparison with other tradable quotas

---

[1] Axel Gosseries is from Fund for Scientific Research and Louvain University (UCL). This chapter builds on earlier work, especially on A. Gosseries and V. Van Steenberghe, 'Pourquoi les marches de permis de polluer? Les enjeux économiques et éthiques de Kyoto', *Regards économiques*, 21 (2004), 1–14. It has benefited from funding from the ARC 09/14–018 Sustainability, the European Science Foundation (ESF) Rights to a Green Future projects, and from a Dyason Fellowship (2013) at the University of Melbourne. I wish to thank especially G. Arrhenius, N. Eyal, D. Halliday, I. Hirose, K. Lippert-Rasmussen, A. Preda, K. Steele and, last but not least, J. Moss for their comments on earlier versions of this chapter.

[2] See e.g., M. Cardwell, *Milk Quotas, European Community and United Kingdom Law* (Oxford: Clarendon Press, 1996); D. Ellerman, P. L. Joskow, R. Schmalensee, J-P. Montero and E. M. Bailey, *Markets for Clean Air: The US Acid Rain Program* (New York: Cambridge University Press, 2000); D. de la Croix and A. Gosseries, 'Population policy through tradable procreation entitlements', *International Economic Review*, 50 (2009), 507–42 (exploring Boulding's 1964 proposal).

systems, be they in place (e.g., tradable fishing quotas) or merely proposed (e.g., exchangeable asylum seekers quotas). I will limit myself to carbon trading in a global context and explore whether there are reasons to be concerned about its fairness.

I will proceed in three steps. First, the three main components of a tradable quotas scheme will be presented. I will explain how the issue of their fairness can be dealt with in the section 'Cap, allocate and trade'. Second, I will briefly look at the advantages of tradability ('What can be gained from tradability?'). Finally, I will explore with some detail possible reasons to object to tradability in the specific case of emission quotas ('What's (possibly) wrong with tradability?'), before concluding.

Cap, allocate and trade

Let me begin with a simplified account locating tradability within a climate regime involving three components: a global cap on emissions for each successive period, a corresponding allocation of quotas to different geographical zones for that period and the tradability of quotas. Any cap-allocate-and-trade scheme designer has to decide first on the total amount of emissions that should be allowed. We then need to distribute the reduction efforts across different emitters, which leads to an allocation of emission entitlements. Finally, a decision has to be taken as to whether such emission entitlements should be tradable.

Insofar as a global cap is concerned, consider that countries agree that humanity as a whole is emitting too much in terms of greenhouse gases (GHGs). Imagine that they come to a more or less explicit agreement on a global cap, e.g., through defining a maximum global 2°C increase compared to a baseline year. A fair level of emissions can be interpreted as one that, while being harmful, would not be considered wrongful. It benefits some and hurts others in ways that are deemed morally acceptable overall, typically because the policy is beneficial in net terms and because winners-to-losers transfers are put in place. In other words, when states agree on a global cap for a given period, they make a decision that should be analysed through the prism of fairness. This point is often neglected. Deciding on a cap involves more than factual claims on dangerosity thresholds based on evidence from natural scientists alone.[3] When we set a global cap, we are in fact deciding on a given level and distribution of harms and

---

[3] Cf. E. Page, 'Cashing in on climate change: political theory and global emissions trading', *Critical Review of International Social and Political Philosophy*, 14/2 (2011), p. 262 ('This is largely a natural scientific enterprise, although one which also involves normative elements such as the interpretation of the amount of climate change that would be 'dangerous' for human life'); S. Caney, 'Two kinds of climate justice: avoiding harm and sharing burdens', *Journal of Political Philosophy*, 21/4 (2014), s. I ('I think we have reason to focus on what would most effectively prevent the onset of dangerous climate change and then consider what responsibilities would follow from that').

benefits, between different regions of the globe, some being more vulnerable or more (ir)responsible than others, and between different generations, some having to face a warmer and more unstable climate than others. And this is only justifiable if it is in line with what justice requires.

Seeing the global cap as a first key normative step leads in turn to at least two considerations. First, we operate globally without any *general* redistributive tax-and-transfer scheme. In domestic contexts where decent tax-and-transfer schemes obtain, general distributivists may want to decide on cap-allocate-and-trade schemes on purely efficiency-based, aggregative grounds, leaving distributive concerns to a general tax-and-transfer regime.[4] In the absence of the latter at the global level, even *general distributivists* – i.e., those who believe that specific regimes (dealing with, e.g., environmental or labour issues) should not address distributive concerns in their design, such concerns having to be addressed through general tax-and-transfer – need to integrate fairness considerations in the design of a global climate scheme.[5]

This leads to a *second* point. One may accept that a climate regime should take fairness issues into account, be it due to the absence of general tax-and-transfer scheme or not. And yet, one could still want to divide up 'labour' within this climate scheme. For instance, the global cap would be set on purely *aggregative* grounds, *distributive* concerns being dealt with at the allocation level only, i.e., when translating the global cap into country-specific quotas. It is unclear though why both aggregative and distributive concerns could not be taken into account at each level. Alternatively, even if one considers distributive concerns as relevant to cap definition as well, one might want to address intergenerational justice concerns exclusively at the cap level, dealing with global justice issues exclusively through selecting the right allocation algorithm between countries. As any cap will have both an intergenerational and a global distributive impact, one should take both dimensions into account at this very first stage too. It is unrealistic to expect requirements of global justice to be met through merely adjusting the allocation of quotas to the countries involved, irrespective of the global cap level.[6]

Let us assume that a fair global cap on emissions obtains. We then have to divide up the total amount of emission rights per period into, e.g., national quotas – this is the scenario envisaged here. There is no way we can do this without having at least a rough idea of the purpose we assign to climate justice. Let us assume that we generally favour a conception of justice that gives a very

---

[4] J. Tobin, 'On limiting the domain of inequality', *Journal of Law and Economics*, 13 (1970), 263–77.
[5] On this: E. A. Posner and D. Weisbach, *Climate Change Justice* (Princeton University Press, 2010), pp. 93 and 97; M. Fleurbaey, 'Justice et climat: alliance ou tension?', *Raison Publique*, April 2010.
[6] We leave aside a third problem associated with the difficulty of setting a fair provision level whenever public goods or bads are at stake.

significant weight to distributive concerns without disregarding efficiency. How does it translate into a theory of climate justice? Consider three approaches:

The purpose of climate justice

Isolationist corrective: climate justice should aim at making sure that people from different regions and generations are not worse off than they would otherwise have been in the absence of human-induced global warming.

Isolationist distributive:[7] climate justice should apply a general distributive principle to the allocation of emission rights, irrespective of the level of other resources that different people and generations enjoy.

Generalist distributive: climate justice should contribute to the promotion of distributive justice in general, all goods considered.

The first approach aims at making sure that human-induced climate change does not worsen what people and generations have – i.e., their respective levels in a general distribution – compared to a 'natural climate' scenario. The idea is to merely neutralise human-induced climate change's impact. It neither cares about correcting natural climate inequalities, nor wants to grant equal rights to emit, nor aims at contributing through this specific scheme to the eradication of unfair inequalities in general. I suspect that a significant amount of climate negotiators hold this 'corrective' view. Let me point here at two unfounded potential 'grounds' for it. First, one might think that since there are no grounds for global distributive justice, we should limit ourselves to a corrective approach. This would not work as there is no reason to aim at correcting departures from a baseline distribution if the latter cannot be justified as being fair in the first place. A global corrective approach necessarily presupposes that we accept some idea of duties of global justice. Second, one might believe that specific policies should aim at specific goals, leaving distributive work to other more general policies. As we said, there are no such general policies in place at the global level. Specific schemes are the only thing we have and we only have very few of them. Let me add to these two points of criticism the idea that if a *distributive* view allocates quotas in a manner that is sensitive to the degree to which various countries have contributed and are affected by climate change, it does incorporate a *corrective* logic from within too.

We may then want to go for a different – slightly more ambitious – view that is explicitly distributive while not going as far as turning a climate regime into a tool for furthering global and intergenerational justice in general. The idea of this intermediary view would be to apply a general principle of justice 'in isolation', regardless of the distribution of other goods. Those attracted to the idea of 'equal

---

[7] This 'isolationist' label originates from S. Caney, 'Just Emissions', *Philosophy and Public Affairs*, 40 (2012), 259 ('Method of isolation'). For a recent discussion of the distinction between a distributive and a corrective understanding of climate justice: Caney, 'Two kinds of climate justice'.

emission rights per capita', independently of the level of natural and non-natural resources of each country, may implicitly be endorsing such an 'isolationist' view. More importantly for us here, 'isolationism' is actually compatible with departures from equality per capita, for instance if our view about justice in general is sensitive to responsibility and historical emissions or to the degree to which the respective countries suffer from global emissions.[8]

However, an isolationist view does not resist close scrutiny as there is no plausible reason to look at things 'in isolation', at least as a first best strategy. Isolationism may actually converge to a certain extent with a generalist distributive approach if it applies to each and every specific policy. But we don't have it in place at the global level in each sector. Then, why insist on equalising resource x between A and B if we know that B is much richer than A with respect to resource y?[9] In that sense, an isolationist approach may not only be inefficient in general. It may also be unfair.

To put it briefly, an 'equal emission rights per capita' position brings us closer neither to a corrective approach (as correction may require unequal rights to compensate for unequal impacts or unequal responsibilities), nor to general distributive fairness (for exactly the same two types of reasons). Political feasibility and rhetoric aside, it seems then that the only tenable position for a global and intergenerational distributivist is to make sure that a climate regime pursues the third purpose above, i.e., the promotion of general distributive justice.[10] A regime applying to a specific good thus needs to take into account the general situation of the countries at stake, including their position in the global distribution of wealth.

Once this is understood, there would be no reason not to allocate more emission quotas to unfairly poor countries. We won't go into details here. But this would definitely justify an unequal redistributive allocation of emission rights, and even more strongly so if such rights are tradable. Understanding the purpose of climate justice in such affirmative, generalist terms – in Tobin's sense – is crucial at the allocation level. While some 'isolationists' may want to go for 'equality per capita', a corrective approach will typically grant extra quotas to countries that are more *affected* by climate change and a generalist approach grant extra quotas to countries that are simply *unfairly poorer* in general, be it for climatic or non-climatic reasons. Proposals that endorse e.g., an inverse to gross domestic product (GDP)/head allocation can be labelled generalist distributivists in this sense. Note that identifying the

---

[8] In the latter case, we would adopt what Caney refers to as a non-atomist isolationist view: S. Caney, 'Just emissions', *Philosophy and Public Affairs*, 40/4 (2013), p. 260.

[9] The same critique applies to a free-riding based approach to historical emissions. See A. Gosseries, 'Historical emissions and free-riding', *Ethical Perspectives*, 11/1 (2004), 36–60.

[10] See A. Gosseries, 'Cosmopolitan luck egalitarianism and the greenhouse effect', *Canadian Journal of Philosophy*, suppl. Vol. 31 (2007), 279–309, pp. 304–6.

purpose of climate justice along those lines is also key for cap definition, not only for the allocation level. For instance, if we could assume that the next generation will be much better off than we are in non-climatic terms, it would affect the level of climate degradation that we would be entitled to allow, and, as a result, the acceptable global cap level.

Consider then tradability, which is the focus of this chapter. Tradable quotas differ both from non-tradable quotas and from Pigovian taxes. Under a non-tradable, fixed quotas regime, a global cap is defined for each period and then decomposed into country-specific quotas that are not tradable. While it does not mean that the allocation should be uniform, it entails that countries are not allowed to emit beyond the limit allowed for by their quota. In contrast, when quotas are tradable, a country that can afford not to use up all its quota (e.g., because it has massively invested in green technology or because its economy has crashed) can sell its unused ones to another country. The latter is then allowed to emit beyond its own quota in the same proportion.

The other alternative consists in approximating the desired global cap through a Pigovian tax. As for the fixed quotas and the tradable quotas scheme, the imposition of a tax limits emissions. Both tradable quotas and a tax system are referred to as incentive-based. The main difference is that, if complied with, a quota system guarantees that a given global level of emissions will not be crossed. No such guarantee obtains with a tax scheme, notwithstanding the fact that a Pigovian regulator will of course try to anticipate which tax rate will lead to the desired total amount of emissions. While the tax fixes the price and lets quantities adjust, the tradable quotas scheme fixes the quantity and lets the price adjust.

While tradable quotas are more flexible than fixed ones, they should not necessarily be seen as tools to deregulate a previously regulated domain. Rather, they are instruments through which regulators introduce new constraints in a cost-effective manner. It is the regulator's intervention that brings these goods into existence and renders emission reductions marketable and valuable. In the absence of the regulator's constraint, emission reductions would be worthless on such a market and the latter would not even exist. While in the domestic case, the state will generally be directly involved in the design of the market, it is more fundamentally its prior intervention through introducing a cap that makes this market meaningful in the first place.

A trickier question is whether the distinction between the three purposes of climate justice presented above is likely to affect our view on tradability. If we anticipate efficiency gains, we could afford to have a lower cap compatible with fairness. But this holds for the three views on the purpose of climate justice. More interestingly, if some purposes require an allocation that departs from equality per capita, tradability might possibly allow for quotas levels that go below subsistence level to less affected, more responsible or more advantaged countries, given the possibility for them to buy quotas back from other – respectively – more affected, less responsible or less advantaged countries.

What can be gained from tradability?

Before looking into possible objections to tradability, it is important to understand the nature and size of its possible benefits.[11] If there were no such benefits, our general conclusion could of course be affected. Let me insist on several benefits associated with tradability: flexibility, efficiency, coordination and information. First, there is a flexibility dimension. Tradability allows some players to emit above their quota level without any wrongful impact on the global environment. Such a flexibility, even if it could not be automatically reinterpreted as an efficiency gain, would as such be a virtue of the scheme, freedom being increased for these actors.[12] This flexibility is, however, only compatible with meeting the global cap if, whenever some emit above their target, others correspondingly emit below their target. This is especially harmless whenever we deal with uniformly mixed pollutants, i.e., pollutants for which the place of emission and possible concentration of sources in some locations does not impact on its global dispersion. Uniform mixing should not be confused with uniform impact. The fact that a pollutant is uniformly spread around the globe does not entail that some regions are not more vulnerable to the effects of global warming than others. One should note as well that uniform mixing, while allowing for the free allocation of sources, also reduces the incentives for emitters to reduce emissions as the local impact of emissions does not tend to be strong.

Second, looking at *efficiency* per se, the scheme plays on two mechanisms. Tradability makes it possible to locate reductions where they are the cheapest (allocation function). Tradability also makes it *desirable* for individual agents to do so, as it generates an incentive to do better than one's target, which is crucial from a dynamic perspective (incentive function). Under a fixed quotas scheme, players are neither allowed, nor willing to locate reductions where they would be much cheaper to implement.

The third and fourth benefits of the scheme have to do with the existence of a market and of a price mechanism associated with it. The market plays a coordination function, allowing for supply and demand to meet. More importantly, however, the price mechanism is crucial for two actors, and especially

---

[11] See e.g., D. Satz, *Why Some Things should Not be for Sale: the Moral Limits of Markets* (Oxford University Press, 2010), part I.
[12] See S. Caney and C. Hepburn, 'Carbon trading: unethical, unjust and ineffective?', *Royal Institute of Philosophy*, Supplement 69 (2011), 201–34, p. 206.

so if we operate under admittedly often unrealistic conditions of perfect competition (atomicity, symmetric information and so on). First, for potential sellers and buyers. When prices go up, potential sellers will want to invest in emission reductions in order to sell part of their quotas. Second, the price mechanism offers very valuable information to the regulator. If prices go down, it means that economic agents don't have such a hard time meeting their environmental targets. The regulating agency can then decide to lower the cap and quotas further at the next period, keeping prices high enough, which preserves the incentives associated with the scheme. Such price-based information, while present in the tax case, is absent in the fixed quotas case. Hence, while it invites duty holders to keep reducing their emissions further, tradability also provides regulators with the information to set the global cap at the right level. A fair cap does not need to remain the same forever. Its fairness also depends on the assessment we make of the size of reduction costs. If these costs turn out to be less significant than expected, including for the least well off, there is no reason why the cap could not be reduced accordingly, given the costs that such a non-reduction would impose on others. And the reverse is true as well.

We have pointed at a set of virtues of tradable quotas (allocation, incentive, coordination and pricing). Can we identify cases in which tradable quotas schemes can only meet part of these functions? If so, we might have to consider objections to such schemes with lesser benefits in the balance. Let me concentrate on two illustrations, one outside the climate context (tradable asylum-seekers quotas) and the other within (project-based offsets in open contexts).

Consider first the tradable asylum seekers quotas proposal – to be distinguished from a tradable refugees quotas scheme, once their refugee status has been granted.[13] The idea is to host asylum seekers in countries where it is cheaper to do so, while remaining safe for them. If we consider potential host countries, once the criteria for being granted asylum are internationally defined, the total amount of asylum seekers will be – at least in part – beyond the pool of host countries' control. The 'cap' is given. It would seem to follow that the cost-revelation-to-the-regulator function would be absent from such a scheme. This would differ from the climate case to the extent that the group of safe countries who can potentially host asylum seekers could not simply collectively decide to host more asylum seekers if it turns out that hosting them is cheaper than expected. At least in theory, they have to host them, regardless of whether handling their applications and hosting them is cheaper than elsewhere.

---

[13] See J. Hathaway and A. Neve, 'Making international refugee law relevant again: a proposal for a collectivized and solution-oriented protection', *Harvard Human Rights Journal*, 10 (1997); P. Schuck, 'Refugee burden-sharing: a modest proposal', *Yale Journal of International Law*, 22 (1997), 243–97.

This is probably too simplistic. If the potential for efficiency were significant, and if there were no significant objection from the perspective of asylum seekers' rights, such a scheme might play a coordination and an allocation function, as well as generate incentives to handle the applications of asylum seekers in more efficient ways. Revealing the costs to the regulator (say, the assembly of safe potential host countries), might actually have some virtues. If the administration and hosting costs are high, it can put extra pressure on collectively addressing the source of the problem, e.g., through more serious diplomatic efforts against regimes that violate basic human rights. If, on the contrary, administration and hosting costs turn out to be low as revealed by the price mechanism, considering more generous asylum criteria might be encouraged. Both of these things should be done anyway. However, the existence of a market, while reducing the freedom of asylum seekers, might actually have this virtue. Whether such a market is preferable to, for e.g., 'burden-sharing' schemes without tradability or no quotas scheme at all will depend on how realistic the gains from having such a market for potential asylum seekers are and on how serious the costs on actual asylum seekers it would entail. What matters to us here is that, in theory, such a scheme could actually exhibit all the functions discussed above, contrary to what could be expected at first sight, which does not admittedly entail that it would be desirable all-things-considered.

Consider now another example, climate related this time: project-based offsets in an open context, i.e., in a context in which not all players committed to a national cap. To put it briefly, an economic actor from a country that committed to a national quota invests in another country where there is no such commitment, in order to achieve emission reductions in that country, and having them counting as domestic emission reductions in the country of origin that committed to a national quota. Imagine that such offset projects also develop through a market. And imagine that we could reasonably guarantee the reliability of such a scheme in terms of additionality. Would it be so that some of the functions identified above would not be met in such an open context? The coordination, allocation and incentive functions are clearly present. However, what about the possibility for the regulator to adjust the cap on the basis of what the prices reveal? We said that one of the countries involved did not have a cap. However, we should bear in mind that the cost-revelation function is still useful for the country that committed to a national quota. If it sees that its economic actors can reduce their emissions more easily than expected, be it through offset, it can consider accepting a more stringent quota.

So, these two examples illustrate the fact that even in contexts with more specific features, all the functions identified above can potentially play a role. This does not mean of course that objections cannot be raised against such schemes. We now turn to them.

What's (possibly) wrong with tradability?

In order to discuss various possible objections to tradability, I will first assume the existence of a fair cap and a fair allocation in a closed system. As we said, a tradable quotas regime is closed whenever all players committed to a national cap. This does not entail that all trades need to take a quota-based form rather than a project-based form. It merely entails that the baseline against which an emission reduction should be assessed is defined through a measurable quantitative commitment. The stronger assumption is that both the cap and the allocation algorithm are fair. This means that if a country remains within these limits, it can be said to be acting in a fair manner. This twofold assumption will be relaxed as we go along, envisaging both a closed system with an unfair cap and/or allocation and the case of offsets in an open system.

Let me address successively a set of possible objections to tradability:
(a) buyers should not get the right to do wrong through buying extra entitlements
(b) sellers should not get paid for what they should do anyway
(c) buyers/sellers lose track of the wrongness of emissions when trade is allowed
(d) buyers should not expect others to do what they should accomplish themselves
(e) sellers should not be allowed to sell their quotas too cheaply

(a) Does tradability necessarily imply a right to (do) wrong?

With the twofold assumption in mind, consider the first objection.[14] It worries about whether making quotas tradable would not amount to granting *buyers* the (legal) right to do (moral) wrong. We will get inspiration from Goodin's analogy about selling religious indulgences and from Sandel's comparison with someone preferring to drop a beer in Grand Canyon and pay a fine than abiding by the law.[15]

Goodin discusses the analogy with religious indulgences in various ways to which it is not possible to do full justice here. Medieval indulgences consisted of paying cash in order to avoid some punishment after one's death. It didn't imply denying that the act that would otherwise justify a punishment was wrong. Moreover, being able to substitute the expected punishment with cash was also premised on committing not to go on with the wrongful action.[16]

---

[14] See as well Caney and Hepburn, 'Carbon trading', pp. 221–3; Page 'Cashing in on climate change', p. 272 ff.
[15] R. Goodin, 'Selling environmental indulgences', in J. Dryzek and D. Schlosberg (eds.), *Debating the Earth* (Oxford University Press, 1994), pp. 237–54; M. Sandel, 'It's immoral to buy the right to pollute', in M. Sandel (ed.), *Public Philosophy: Essays on Morality in Politics* (Cambridge, MA: Harvard University Press, 2005), pp. 93–6, p. 94.
[16] Cf. C. Spash, 'The brave new world of carbon trading', *New Political Economy*, 15/2 (2010), p. 188.

Nevertheless, allowing some people to buy their way out of standard punishment simply because they are richer, is problematic towards those who cannot afford it (objection to buying).

Sandel's story invites us to question whether doing something wrong and paying the fine should be seen as morally equivalent to not doing that wrong thing in the first place. Sandel asks 'Would there be nothing wrong in his treating the fine as if it were simply an expensive dumping charge?'[17] Sandel's dumping example is premised on the assumption that society considers that beer cans should not be dumped in Grand Canyon in the same way as we drop our rubbish bags on a weekly basis before the local waste collector comes and collect them.

There are serious dis-analogies between these cases that become clear once we retranslate the features of each case in the language of a deadly bullet example.

Dumping a can (Sandel)
I shoot at a person. The person is badly hurt. I pay a fine for it. I consider that this fine is equivalent to a charge aiming at compensating the person.

Buying indulgences (Goodin)
I shoot at a person. The person is badly hurt. I pay a fine for not being punished. I don't deny I was wrong to shoot at that person. I commit to never shoot at anyone again.

Buying extra permits
I shoot at a person. Before doing so, I pay someone to install a bulletproof device between me and the person. The person is not hurt. I deny that shooting was wrong under those conditions.

A crucial difference between the dumping and the extra permits cases is that the latter case is equivalent to paying someone to immediately collect my beer can on the floor. No wrong would follow. Similarly, contrary to the tradable permits case, in the indulgences case, the cash transfer does not fund actions aimed at preventing the wrongful effects to take place. The sole function of cash transfers in the medieval indulgences case is to shift from a sanction in kind to a cash and recognition fine. This is very different from a situation in which one would shift from an effort in kind to a cash effort while guaranteeing that the effects will remain *exactly the same* and that nobody will suffer any wrongful consequences. In carbon trade, one is buying the right to engage in a practice that, were the counterpart agent not engaging in corresponding emission reductions, would otherwise be wrong.

So, in the carbon trading case, we don't buy the right to do wrong and we don't buy the right not to be punished for doing wrong or for having done

---

[17] Sandel, 'It's immoral to buy the right to pollute', p. 94.

wrong. We buy the right to do something that will not be wrongful because of the corresponding extra duty it entails on others – against payment – to keep or reduce their own emissions below the quota that they would otherwise be entitled to use. If this corresponding duty to lower emissions further is complied with on the seller's side, no difference in effect follows, compared to the fixed quota scenario.

Let us mention here the second challenge. Goodin's indulgences and Sandel's beer example differ quite significantly from the tradable quotas case. One may then be tempted to conclude that, provided that we operate on a closed system with fair cap and allocation, there would be nothing wrong towards potential victims of climate change with some players *buying* extra quotas in exchange for others to reduce their emissions. The related issue is whether there is something wrong for sellers to expect being paid for doing something that they *should* do anyway. Here again, one could stick to the view that if sellers act within the limits of a fair quota, they *should not* reduce their emissions further. They can do so if they want, including in exchange for payment. But they should not. This answers the second worry before (b) the seller does not get paid for what he should do anyway. He gets paid to remain below what he would otherwise be (fairly) entitled to do, given our twofold assumption of a closed system with fair cap and allocation. Of course, if we relax the 'fair cap and allocation' assumption, the story becomes different and this second objection would hold.

(c)   Does tradability remove the moral stigma that should be attached to emission rights?

Should we also reject Sandel's intuition that 'turning pollution into a commodity to be bought and sold *removes the moral stigma* that is properly associated with it'?[18] At first sight, this intuition might make sense in a world in which we relax our fair cap/allocation assumption since acting legally will not necessarily entail acting morally. After all, in the real world, the cap and/or the allocation are (likely to be) significantly unfair, especially if we adopt a generalist distributivist interpretation climate justice. Keep in mind though that the question is whether tradability *adds* an extra moral problem to the one raised by cap/allocation unfairness.

Sandel's 'moral stigma removal' worry may be looked at statically or dynamically. The most relevant perspective here is the dynamic one, in which tradability has an impact on the total amount of emissions, under an 'unfair cap/allocation' scenario. There is no doubt that tradability can make reductions cheaper and that this should allow negotiators to bring the legal levels of emissions closer to the fair levels. What is less clear is the relative

---

[18] Sandel, 'It's immoral to buy the right to pollute', our italics.

importance of the moral stigma message that such tradability may convey, and of its effect on people's willingness to reduce emissions further. In theory, if tradability were to send a message to emitters such that they may become totally relaxed about emissions, this may reduce the pressure towards a fair solution that could be such that we would end up with a less fair climate scheme, notwithstanding the cost reduction potential of tradability. The key question here is whether making x tradable sends the message that x is less problematic than if it were non-tradable. If this net 'de-stigmatising' effect were very significant, to the point of being stronger than the cost-reduction potential generated by tradability, there might be a case again tradability.

Should we expect such a net 'de-stigmatising' effect in a realistic 'unfair cap/ allocation' setting? Consider a comparison with two legal regimes, one under which tobacco can both be smoked and sold, and another one under which marijuana can only be smoked, its trade being illegal. It is plausible to believe that the message sent by the legislator is that smoking marijuana is more problematic than smoking tobacco. We may then want to draw the implication that tradable emission quotas are more de-stigmatising than fixed emissions quotas. My view is that this conclusion does not follow whenever tradability is part of a scheme aimed at reducing the size of a problem (overemission, overproduction, overpopulation and so on). In such cases, tradability is meant to make reductions cheaper through maximising the use of low abatement cost solutions. It is plausible to believe that the stronger the need to encourage such reductions, the more problematic non-reduction is meant to be. Tradability may thus be sending the reverse message in the tradable quotas regime than in the tobacco/marijuana case. While Sandel's intuition holds for ordinary consumption goods or practices, this suggests that it does not for goods that are part of a quantitative target aimed at reducing the size of a collective problem. At the very least, GHG quota's tradability sends a mixed message, both de-stigmatising and stigmatising.

The morally stigmatising message associated with tradability may even operate under the stronger 'fair cap/allocation' assumption. If people are granted incentives to go beyond what their entitlements allow, it can be interpreted as a way for the legal regime to express the idea that going beyond one's quota could be morally desirable. And if it is morally desirable, it may be because there remains something morally problematic even with a 'fair' quota. Referring to a regime as 'fair' does not prevent it from being a lesser evil.[19]

---

[19] See J. Aldred, 'The ethics of emissions trading', *New Political Economy*, 17/3 (2012), 339–60, p. 343 ('On the prima facie interpretation of damage, then, every unit of emissions does damage no matter what the level of the cap is. This is the key objection to emissions trading behind the commodification argument') and p. 344 ('Thus, both the weak and strong versions insist that we recognise the damage done by each act of emitting carbon. It is precisely these damages which carbon markets fail to recognise. In such markets, emissions below the cap are not merely "not morally wrong", but legitimate entitlements for buyers and sellers'). See as well Goodin, 'Selling environmental indulgences' (distinguishing 'the right thing on balance' and a 'wrong').

If so, it would still be morally good if a player who can do so were to reduce its emissions beyond what is fairly and legally required. The exact reason why a fair cap is necessarily a lesser evil is probably related in part to the fact that our emissions are uniformly mixed, which can probably be accounted for as a specific form of public bad, i.e., a pollution problem that involves the non-excludability of victims with the specific complication that the pollutant spreads uniformly and regardless of the location of sources. However, the fact that it is a *public bad* of a special kind only leads to a problem if victims are not being compensated. But if they are not, we fall back onto an 'unfair regime' scenario. Note as well that there is another difference in the nature of the tobacco/marijuana and of the emission rights examples. When you allow people to sell tobacco, you allow them for more than selling the right to smoke. It is as if in the climate case you were to add to selling the right to emit GHGs a right to sell GHG emitting devices.

Let me add that when Sandel asks 'Should we buy the right to pollute?', the use of the word 'right' is potentially as de-stigmatising as the word 'buy'. However, even here, the implications are unclear. Consider the legal right to smoke (with risks for passive smokers) or to drive (with the risks for pedestrians or those with sensitive lungs). People know that having a right does not make the exercise of this right harmless for others. And that even remaining within the realm of their right (e.g., complying scrupulously with the speed limits) does not mean that others won't be at risk.

We have shown that tradability is not necessarily a feature that removes the moral stigma that should be attached to unfair emissions. We have actually insisted on the fact that tradability might even send a stigmatising message to those who properly understand the logic of a tradable quotas scheme. Moreover, Sandel's worry should only concern us if the net expressive effect of tradability were a de-stigmatising one (it is likely to be mixed at best) and if it were sufficiently strong to counterbalance the benefits of cheaper emissions allowing for a more stringent cap or a fairer allocation (which is hard to argue).

(d)   The 'do-it-yourself' objection

In a certain sense, the two previous sections discussed objections that, if conclusive, would also imply that it is better for potential buyers to do it themselves rather than to pay others to do it for them.[20] Are there other reasons to expect people to do things themselves instead of paying others? There are certain things that others are simply *unable* to do for us (e.g., sleeping). For other things, we also find it *preferable* to limit the possibility for people to contract others to do it for

---

[20] See J. Tronto, 'The "Nanny" question in feminism', *Hypatia*, 17 (2002), 34–51; Caney and Hepburn, 'Carbon trading', p. 214 ff.

them (e.g., sexual intercourse within a marriage relationship). We may (want to) make the exercise of certain tasks compulsory (voting, maternal leave, school attendance, military service (pacifist case), jury membership, visit to hospitals with victims of car accident for dangerous drivers, slaughterhouse duty for meat consumers (vegetarian case), home tasks (feminist case), burying waste at home (environmentalist case) and so on.[21]

The underlying idea is that it makes a difference whether people do something themselves or whether they pay someone else to do it, which points at a possible non-substitutability of cash and kind. What difference does it make exactly? In some cases, it is a matter of social cohesion (we want people from all sorts of backgrounds to participate in a given activity, out of fear for social segregation). We also find it important that people actually see with their own eyes the impacts of their choices that they may not want to see. This is what underlies in full or in part e.g., military service, car drivers' compulsory visits to hospitals, slaughterhouse duty and so on. Some of this probably also underlies the concern of some opponents to tradability in the present case too. In fact, the point is both epistemic (knowing what it really entails) and motivational (if I actually need to do it, I'll be politically motivated to act). The same idea is present in one of the justification of the so-called 'all affected principle' in democratic theory. The fact that I have been and that I will be subjecting myself to the decisions to which I contribute both gives me first hand experience about its effects (epistemic) and is such that I will be concerned about such effects when I contribute to future decisions (motivational).

In the climate debate, one may ask: what difference does it make to put the effort in cash (buy extra quotas) or in kind (achieve oneself the reductions at home)? Is there a case against tradability for such reasons, despite the efficiency gains that can be expected from tradability and more generally from some degree of division of labour? Are there good reasons to expect large emitters to achieve a significant amount of reductions at home? One can identify several such reasons. First, if large emitters are also those that are technologically the most advanced, forcing them to make further reductions will trigger technological innovation that will in turn benefit all. Second, it may also matter that those who are involved in negotiations experience how difficult it can be to reduce emissions, to properly assess the efforts of others that might be willing to commit to reduce emissions. Similarly, the fact of experiencing it as a burden more than in the 'far from the eyes' case will generate an incentive to keep being active at the political level to try and fix the problem. So, there is an epistemic case (making people aware of what it means to reduce emissions)

---

[21] See J. Suk, 'From antidiscrimination to equality: stereotypes and the life cycle in the United States and Europe', *American Journal of Comparative Law*, 60 (2012), 75–98.

and a motivational one against full tradability (that people if they need to put in an effort that is non-cash, will take it more seriously politically). However, this presupposes in many cases some non-substitutability of cash and kind. An effort in cash would not be deemed equivalent – epistemically and motivationally – to one in kind.

What should we think about this 'do it yourself' objection? I would say three things. First, the 'do it yourself' case is weaker here than in the prohibition on solid waste export case for instance because of the uniformly mixed nature of at least some of the GHGs. Where you emit and where you reduce does not make any difference as to who will be impacted. Of course, how much you do and how much we all do will make a difference. Second, the 'do-it-yourself' objection provides at best a case against full tradability, not one against partial tradability. Third, it is unclear whether it follows that it is morally compulsory to do it oneself or whether it would be desirable *ceteris paribus*. Efficiency gains from differences in marginal abatement costs should be taken into account in this respect. In short, while for certain things I am convinced that doing it oneself would make/makes a huge difference in people's perception of the problem (e.g., slaughterhouse duty), I don't think that it provides a case to justify banning tradability altogether, be it in the standard cases (e.g., slaughterhouse duty would be epistemically and motivationally effective even if one is not expected to slaughter all the animals one plans to eat) or in the present one.

(e)   Unfairly cheap?

Consider now a last objection.[22] Very often, the concern with tradability is with potential buyers who could not afford to buy because things would turn out being *too expensive*. Here, the problem is different and there are two different ways in which tradability may be problematic because it would lead to *too cheap* deals. The first version is the 'low-hanging fruits' objection.[23] The idea is that early reductions are cheaper to make than later ones, the marginal cost of emission reduction increasing as we go. If early reduction opportunities are sold too cheaply to buyers, selling countries will be left with the most expensive ones.

There are complications with this low-hanging fruit argument. First, if quotas are limited to periods of ten to twenty years, selling emission quotas is not comparable to selling one's best arable land to a foreign country forever at too cheap a price. However, even if the entitlements are limited to a relatively short period of time, it may deprive locals from significant opportunities, often against their will if decision-makers don't have their preferences aligned with

---

[22] Compare Caney and Hepburn, 'Carbon trading', p. 217 ff.
[23] See e.g., U. Narain and K. van't Veld, 'The clean development mechanism's low-hanging fruit problem: when might it arise, and how might it be solved?', *Environmental Resource Economics*, 40 (2008), 445–65.

the interests of the people they represent. This being said, while there is some realism in the low-hanging fruit worry, I would not say that it would justify a case against tradability altogether. It could justify restricting tradability to luxury emissions. Of course if climate emergency circumstances are such that a fair cap would equal the global subsistence emissions level and that the allocation would reflect that by granting no (rich) country with a quota below what subsistence requires and no (poor) country with a quota above what subsistence requires, our concern for guaranteeing the right to subsistence emissions would entail the need for a full ban on trade.

The 'low-hanging fruit' version of the 'too cheap' claim should be distinguished from a second version, i.e., the 'early action' one, that focuses especially on our duties towards future generations. If we think that early action matters, keeping the prices of quotas high will preserve a strong incentive to reduce emissions further. We mentioned earlier that making part of the reductions compulsory at home will trigger technological innovation and path change in technologically advanced countries. The argument differs here and focuses on price, with the same concern for taking path-dependency into account and changing path as quickly as possible.

Neither of these two arguments are strong enough to justify a full ban on tradability. In the low-hanging fruit case, once sellers are properly informed, it may still be better for poor people to sell at too cheap a price than not to sell at all. The undemocratic nature of a country may raise tricky questions of course. But I doubt that we would be ready to prohibit e.g., rice exports from an undemocratic country out of concern for its poor producers if we feared that the regime would take away too much money in taxes. As to the early action argument, it rather provides a case for lowering the cap rather than prohibiting the trade. As a second best, one may want to limit tradability to only part of the quota. But restricting tradability might – dynamically – render agreement on a demanding scheme trickier and delay early action further.

Conclusion

We have insisted on the need to locate 'tradability' within the architecture of tradable quotas schemes and to focus on whether tradability *adds* an extra moral problem to the ones that can be raised by cap and/or allocation unfairness. In this respect, we have seen that the purpose we assign to climate justice matters and that tradability may make a difference here too. After reviewing some of the virtues of tradability, we went through five possible objections. None of them would justify a full ban on tradability. The first one we discussed is whether tradability would imply a right to do wrong. We showed how different the tradable emission quotas case is from Goodin's selling indulgences and Sandel's beer can examples. We indicated why, provided that the cap and the allocation are fair, the objection does not hold. The same holds for a second worry, i.e., the one that consists of saying that tradability leads to a situation in which actors are paid for what they should do anyway. Again, this objection does not hold if the cap and allocation are fair, which suggests that the problem would not come from tradability as such, but from the unfairness of the cap and/or allocation in the first place.

In scrutinising a third objection, labelled 'moral stigma removal' (Sandel), we insisted on the fact that, interestingly enough, tradability in the present case – in contrast with the marijuana case – may actually be sending the message that further reductions need to be encouraged, which implies that their level is not unproblematic. As to the fourth objection, the 'do it yourself' one, it provides at best a case against full tradability. And the same holds for the two versions of the 'too cheap sale' objection. They justify at best the need for a non-tradable portion of each country's quotas.

In the end, there is no serious fairness-based case against any tradability in an emissions quota scheme of this type. This does not mean of course that we may not have good reasons to object to tradability for other goods and services, or that there would not be good reason to object to a tradable emission quotas scheme that would involve an unfair cap and/or allocation. While full prohibition on tradability would not be justifiable, restrictions on full tradability might be justified on grounds of the two last objections, the 'do it yourself' and the 'too cheap sale' ones. While renouncing tradability altogether seems unjustified – unless we find ourselves in the specific and not totally unrealistic emergency circumstances identified in the section 'The do it yourself objection' – trade restrictions might be defended for reasons identified in the 'The do it yourself objection' and the 'Unfairly cheap' sections.

6 A just distribution of climate burdens and benefits: a luck egalitarian view

Kasper Lippert-Rasmussen

Introduction

Many political philosophers who address the issue of a just distribution of the climate burden make something like the following worrying observation: 'the long-term impact of climate change will have predominantly, if not uniformly, adverse impacts on the health, cultural life, and economic prosperity of future human populations'.[1] What this means is that, *ceteris paribus*, members of future generations will be much worse off with global warming than would the (almost certainly different) members of future generations that would exist without global warming.[2] Given certain additional assumptions, this is a weighty reason for the present generation to curb its emissions of greenhouse gasses (GHGs).[3] In fact, something like this is the main reason we should do so. However, it is not a luck egalitarian reason. It is a reason that applies even if future generations will be much better off than us and despite the fact that the inequality that obtains between them and us will be even greater if we limit our emissions of GHGs.[4]

---

[1] E. Page, 'Intergenerational justice and climate change', *Political Theory*, 47 (1999), 53–66, p. 54. Some of these bad effects are effects of climate *change*, as opposed to a *worsening* of the global climate, and, thus, may be transitional – see A. Gosseries, 'Cosmopolitan luck egalitarianism and climate change', *Canadian Journal of Philosophy*, suppl. vol. 31 (2005), 279–309

[2] In this chapter I ignore the complex non-identity problem as my arguments in this chapter are largely unaffected by it.

[3] One additional assumption is that our investing in a reduction of GHG emissions is not worse for future generations, because our investing these resources differently will leave future generations better off in other ways, e.g., in terms of 'productive capacity' (B. Barry, *Democracy, Power and Justice* (Oxford: Clarendon Press, 1989), p. 519) or resource-based 'opportunities' (E. Page, 'Intergenerational justice of what: welfare, resources or capabilities?', *Environmental Politics*, 16 (2007), 453–69, p. 457) that outweighs the harms they will suffer from a warmer climate.

[4] It is virtually certain that global warming is taking place and that continued emissions of GHGs will result in further global warming. However, it is less certain whether global warming will trigger threshold effects such as the shutting down of the Ocean Conveyor and how future technologies will help future generations to cope with or counteract the effects of global warming. It is a huge issue what luck egalitarian justice requires in situations involving risk and uncertainty. However, it is one that I shall largely ignore. Instead, I shall focus on idealised situations where we know what the results of different climate policies will be.

According to luck egalitarianism it is in itself unjust if how well off people are relative to one another reflects anything other than their exercises of responsibility. On this view a distribution can be unjust in two ways.[5] First, it is unjust if some are equally well off despite having exercised different degrees of responsibility. Second, it is unjust if some are worse off than others despite the fact that they have not exercised their responsibility any differently from those who are better off. Some believe that luck egalitarianism, properly construed, does not imply the first claim.[6] I believe it does. To avoid this controversy, which appears to be peripheral to climate change, I shall simply focus on the second way in which the distribution of the burdens and benefits of climate change may result in luck egalitarian injustice.

Climate change *per se* need not result in some being worse off than others in a way that does not reflect their differential exercise of responsibility. We can even hypothesise a world in which global warming results in a reduction of unjust inequality so construed. Imagine a world consisting of two continents separated by a non-crossable ocean, where one continent has a harsh arctic climate and the other a pleasant temperate one and where the former climate, but not the latter, will improve as a result of global warming. In this world boosting the greenhouse effect might reduce unjust inequality between people living on the two continents. However, while global warming *per se* is not proscribed by luck egalitarianism, global warming, as we know it, might be.

First, if global warming leads to future catastrophic climate changes, it might increase unjust inequality between future generations and the (immediate past one and) present ones in that future generations end up much worse off than us and much worse off than our immediate descendants.[7] The Stern report warns that: 'Our actions over the coming few decades could create risks of major disruption to economic and social activity, later in this century and in the next, on a scale similar to those associated with the great wars and the economic depression of the first half of the 20th century'.[8] Call this scenario, where, as a

---

[5] G. A. Cohen, 'Luck and equality', *Philosophy and Phenomenological Research*, 72 (2006), 439–46, p. 444.

[6] S. Segall, 'Why egalitarians should not care about equality', *Ethical Theory and Moral Practice*, 15 (2012), 507–19.

[7] Some readers might think that while inequalities between contemporaries can be unjust, inequalities between distant non-contemporaries cannot. I address this view in the section 'States and generations', see also I. Hirose, *Egalitarianism* (London: Routledge, 2014), ch. 3.

[8] What if the changes are so catastrophic that the Earth becomes uninhabitable? Call this the *doomsday scenario*. In that scenario there will be fewer future generations than there would otherwise be. This would be morally extremely undesirable, but because luck egalitarians are concerned only with unjust inequalities between people who actually exist, the relevant scenario need not be bad from a luck egalitarian point of view. It might even be better from the point of view of equality, if the future generations that could have come into existence had been much better off than all predecessor generations. This shows in a particularly perspicuous way why luck egalitarianism does not capture the most important moral concern in relation to climate change.

result of global warming, future generations end up worse off than the present one, *the disaster scenario*.

Second, global warming might increase unjust inequality between different individuals living within the same generation, now or later. This might be the case if, given the absence of counteracting measures or unintended, counteracting effects, global warming affects different parts of the world in a way that tends to boost already existing unjust inequalities, e.g., if it results in minor changes in temperatures and precipitation in wealthy areas of the world with a temperate climate, but leads to damaging effects on poor and arid areas of the world with a subtropical climate. As a matter of fact, there is some evidence to the effect that the negative impact of climate change 'will be heaviest in the tropical and subtropical regions (where most of the [less developed countries] are), and lighter in the temperate regions (where most of the richer countries are)'.[9]

The fact that climate change may be bad in these two ways means that, strictly speaking, we cannot address the justice of climate change using a two-stage procedure where we first determine how much costs each generation should bear to adapt to or mitigate the effects of climate change and then, within the constraints set by the most equal distribution of burdens across generations, within each generation determine which is the just intragenerational distribution.[10] If the distribution that minimises unjust intergenerational inequalities also is a distribution that involves the greatest unjust intragenerational inequalities, possibly, we should accept a greater extent of intergenerational inequality to reduce the extent of unjust intragenerational inequality.[11]

As indicated, the fact that global warming could significantly reduce the level of well-being of future generations is a good reason for us to reduce our emissions of GHG, although this is not a fact that luck egalitarians care about *qua* luck egalitarians. However, since luck egalitarians can acknowledge values other than equality – indeed, it would be an extreme view that held that *only* equality has value – luck egalitarians may acknowledge this reason.[12] In short, luck egalitarians might be pluralists who acknowledge that global warming is morally undesirable in terms of values other than equality.[13] However, they will insist that luck egalitarianism identifies *one* reason, among others, why climate

---

[9] S. M. Gardiner, 'Ethics and global climate change', in S. M. Gardiner, S. Caney and H. Shue (eds.), *Climate Ethics: Essential Readings* (Oxford University Press, 2010) pp. 3–35, p. 9.

[10] cf. Gosseries 'Cosmopolitan luck egalitarianism and climate change', p. 284.

[11] There are many different ways of assessing the overall amount of unjust inequality (cf. L. S Temkin, 'Justice and equality: some questions about scope', *Social Philosophy and Policy*, 12 (1995), 72–104).

[12] If equality is the only value, it follows that a world in which seven billion people live lives that are equally worse than not existing is better than a world in which seven billion people live great lives, but some people live lives that are slightly better than that of other people.

[13] D. Parfit, 'Equality and priority', in A. Mason (ed.), *Ideals of Equality* (Oxford: Blackwell, 1998) pp. 1–20.

warming is, or at least could be, morally undesirable or, as noted above, even morally desirable. This is the claim I shall focus on in this chapter.

First, I explore some implications of luck egalitarianism in relation to climate. My main claim is that luck egalitarianism offers a plausible account of at least one aspect of how global warming may be unjust (see the following four sections). I shall pay much attention to the issue of responsibility in relation to climate change. I submit that we should revise the way in which many luck egalitarians think about responsibility such that a generation can be responsible in the relevant sense for the adverse climatic conditions that it faces even if these are solely the result of the choices of past generations. I will then show how luck egalitarianism relates to two competing views advanced in the literature about what a just distribution of burdens and benefits of climate change should look like (see sections 'Polluter pays principle' and 'Miller's principle of equal sacrifice'). One important claim forwarded in this chapter is that luck egalitarianism does not capture the main reason why we should make intergenerational sacrifices to reduce global warming and, from a luck egalitarian perspective, global warming, as we know it, could even be a good thing!

Does it matter if climate change is man-made?

Most scientists agree that present climate changes are to some extent, if not almost exclusively, the result of human activities. However, some believe that other causal factors play a significant role and a few scientists have argued that past climate changes are to a large extent the result of natural factors, e.g., natural variations in the sun's activity.[14] This is a debate that philosophers as such cannot take part in. However, on some views of justice, it makes a big difference, morally speaking, which of these views is true.[15] For instance, if we believe that the polluter pays principle (to which I return in the section 'Polluter pays principle') is the principle of distributive justice that governs how the benefits and burdens of global warming should be distributed, it is crucial. For on this view there is no climate burden to distribute if climate changes are wholly non-anthropogenic.[16]

---

[14] S. Conor, 'Sun sets on sceptics' case against climate change', *The Independent*, Monday, 14 December 2009. Retrieved from www.independent.co.uk/environment/climate-change/sun-sets-on-sceptics-case-against-climate-change-1839875.html.

[15] S. Caney, 'Climate change, human rights, and moral thresholds', in S. M. Gardiner, S. Caney and H. Shue (eds.), *Climate Ethics: Essential Readings* (Oxford University Press, 2010), pp. 163–77.

[16] As a matter of fact, the polluter pays principle is more complicated to apply than it is often assumed. Suppose, counterfactually, that in the absence of GHG emissions the Earth would have cooled down, and that this would have had a greater adverse impact on people than GHG emissions-induced global warming. In this case, it is unclear that those who emit GHGs and thereby are causally responsible for the adverse effects of global warming have *any* moral obligations to bear the climate burden. After all, we would have been worse off in the absence of GHG emissions.

On the luck egalitarian view, we can set aside this issue as having no fundamental significance from the point of view of justice. On this view, there is no difference regarding distributive injustice between someone being worse off as a result of the actions of other individuals which were not a result of their choice or fault and someone being worse off as a result of natural causes which were not a result of their choice or fault.[17] In both cases someone ends up worse off than others in a way that does not reflect any differential exercise of responsibility. This is the morally relevant fact that makes inequality unjust and this fact is present whether global warming results from human activities or not. This point is a rather obvious implication of luck egalitarianism, but even so it is a quite significant point given that it enables us to set aside completely a discussion which, at least some years ago, was conducted with some intensity among climate scientists.[18]

Intergenerational luck egalitarian justice and the reduced growth scenario

As indicated, luck egalitarianism may not see anything bad about global warming from the point of view of equality, even if it leads to a future climate that is worse than it would have been in its absence. In fact, if we extrapolate from the previous 250 years of global economic growth, future generations will have better technologies and greater wealth than us and, to an even greater extent, better technologies and greater wealth than our predecessors.[19] On the assumption that this growth, albeit perhaps at a significantly slower rate, will continue even if global temperatures rise considerably – say, because future generations will benefit from new technologies though some, but not all, of the resulting gains will be offset by a worse climate – our using resources to reduce GHG emissions for the benefit of future generations will amount to worse off people giving up some of their resources for the benefit of better off people. Call this *the reduced growth scenario*. One reason why it might be more realistic in terms of what will happen in the next generation or two than the disaster or the doomsday scenarios mentioned in the introduction to this chapter is that, presently, economic growth

---

[17] Temkin, 'Justice and equality: some questions about scope'. Cf. J. Rawls, *A Theory of Justice* (Cambridge, MA: Harvard University Press, 1971).

[18] This claim needs to be slightly qualified in the light of the revised formulation of luck egalitarianism that I propose in the section 'Responsibility'. However, the crucial claim here, i.e., that even if climate change is non-anthropogenic the resulting extra inequality is unjust, is unaffected by this revision.

[19] Stern assumes that 'the world economy is likely to be perhaps three times bigger in mid-century than it is now'. N. Stern, 'The economics of climate change', in S. M. Gardiner, S. Caney and H. Shue (eds.), *Climate Ethics: Essential Readings* (Oxford University Press, 2010), pp. 39–86, p. 45. Admittedly, not all of our descendants will have greater income than all of their predecessors and there are complicated issues about how to weigh the fact that the *proportion* of poor people falls against the fact that the *number* of poor people increases in relation to the degree of unjust inequality.

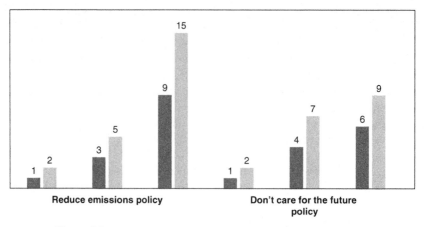

Figure 6.1

is tied to increased GHG emissions such that if the future will see no economic growth, presumably, GHG emissions will have stabilised themselves at the present level (which, admittedly, still will involve rising concentrations of GHGs in the atmosphere and, thus, further global warming).[20]

For simplicity, assume that world history consists of three generations: a pre-GHG generation, the present GHG-emitting generation and the future generation. We can do nothing about the past, but we, i.e., the present generation, can choose between a 'Reduce emissions' policy – the first three pairs of columns in Figure 6.1 – and a 'Don't care for the future' policy – the last three pairs of columns in Figure 6.1. The left-hand column in each pair of columns represents the worst-off within the relevant generation, the right-hand column the best-off and the height of the columns represents how well off they are in terms of whatever is the relevant currency of distributive justice, e.g., resources, welfare or capabilities. This gives us two scenarios:[21]

On the assumption that none of the generations involved is responsible for the global climate during the time at which they exist – I'll return to this assumption in 'Responsibilities' – the 'Reduce emissions' policy seems clearly worse from the point of view of equality. This is not to say that this policy is not best all things considered. Pluralist luck egalitarians may well think that the present generation

---

[20] Note also that much of the present growth in GHG emissions reflects economic growth in poor countries like China and India, arguably, as a result of which global inequality is reduced.

[21] Admittedly these two scenarios are very simplistic and unrealistic, e.g., no one lives a life worse than not existing; in the 'Don't care for the future' scenario, no future person will be as badly off as the worst-off members of the current generation; and in the 'Reduce emissions' scenario, every future person will be better off than the best off members of the current generation. However, for present presentational purpose these features are irrelevant.

is morally required, all things considered, to bear these costs of reducing emissions, e.g., because the sum of the costs for the present generations, i.e., three, is smaller than the sum of benefits enjoyed as a consequence thereof by the future generation, i.e., nine, and that this gain morally outweighs the loss in terms of increased inequality. However, such a moral requirement to reduce emissions is not then grounded in luck egalitarian justice *per se*.[22]

It might be replied that it is implausible to hold that the 'Reduce emissions' policy is morally bad in *any* way. However, this response is untenable. First, reductions impose burdens on the present generation and, surely, it is bad that the present generation is worse off than it would have been had the reductions not been made. Second, whichever of the two policies is adopted the future generation will be better off than the two previous ones. Hence, it is not as if the 'Don't care for the future' policy will violate, say, a plausible Lockean clause to the effect that one generation should leave 'enough and as good' for succeeding generations.[23] Third, if we adopt the 'Reduce emissions' policy this will mean that worst off members of the present generation will make sacrifices for the benefit of members of the future generation, all of whom will be better off than they themselves are. However, it is not clear how distributive justice can require such a sacrifice. Certainly, hardly anyone thinks that, setting aside entitlement, desert etc., intragenerational distributive justice requires that worse off people make sacrifices for the benefit of better off people. Why should intergenerational justice be any different?[24]

In claiming that, in the reduced growth scenario, it is not desirable from the point of view of equality that the present generation bears costs for the benefit of the future generation, I assume that climate is just one component in the overall package with respect to which people's relative positions should be measured. In principle, luck egalitarians could hold, Michael Walzer-style, that climate is a particular sphere of justice of which it is true that it is unjust if, say, one generation is worse off than a previous one climate-wise.[25] I find this position very unattractive. There already *are* and always have been great climate inequalities between people. The climate in Oslo is worse than the

---

[22] Similarly, prioritarians who give great weight to benefits to the worse off may favour the 'Don't care for the future' policy, while prioritarians who give less weight to benefits to the worse off may favour the 'Reduce emissions' policy.

[23] Page, 'Intergenerational justice and climate change', p. 55.

[24] Note that Henry Shue's sufficientarian principle – H. Shue, 'Global environment and international inequality', in S. M Gardiner *et al.* (eds.), *Climate Ethics: Essential Readings* (Oxford University Press, 2010) pp. 101–11, p. 108 – that benefits and burdens should be shared so as to make sure that everyone has enough for a 'decent human life', which he applies to an intragenerational setting, applied to an intergenerational setting might imply – especially if the threshold at which a human life is decent is set at a high level – that future generations should bear most of the climate burden.

[25] A. Dobson, *Citizenship and Environment* (Oxford University Press, 2003), p. 101ff; M. Walzer, *Spheres of Justice* (Oxford: Basil Blackwell 1983).

climate of Dubrovnik.[26] Yet, to my knowledge no egalitarian has claimed that such climate inequalities *per se* are unjust even if offset by inequalities in other spheres such as the fact that people in Oslo are wealthier than people in Dubrovnik.[27] Similarly, when egalitarians point to ways in which global warming is unjust they often note that the worst climate effects will fall on those who are already worse off in other respects than climate. This suggests that for most egalitarians, suddenly to appeal to the separate sphere of climate justice view in relation to the present line of argument is an ad hoc move whose motivation is simply to avoid what they see as an unwelcome implication of their favoured view when it is applied to an intergenerational setting.[28]

Another way of trying to avoid the luck egalitarian conclusion with regard to the reduced growth scenario would be to point out that although the height of the columns in Figure 6.1 represents whatever is the relevant currency of distributive justice my argument presupposes that resources are that in terms of which we should be equally well off. After all, the relevance of economic growth is that we end up having more resources available. However, another plausible view is that we should be equally well off in terms of welfare. If, as some might think, future generations will not have higher welfare simply because they will have much more resources available than we have, and because worsening of the global climate will reduce – let us suppose – their level of well-being, we – members of the present generation – should make sacrifices for the benefit of future generations to avoid unjust inequality in favour of the present generation.

This line of argument strikes me as more promising than the previous one. I am inclined to think that welfare is that, or at least one component in that, in terms of which people's distributive shares should be compared.[29] However, I am sceptical of the view that more resources do not generally translate into more welfare. To be sure, there are uses of resources of which this is not true. But it is equally true that there are other uses that clearly increase our welfare,

---

[26] It might be objected to my use of this example that while climate change induced inequalities are man-made, the climate inequality between Oslo and Dubrovnik is not man-made. In response, I note that Walzer's spheres of justice account is insensitive to this difference as such. Also, from a luck egalitarian point of view the difference noted is irrelevant provided that even in the cases of man-made climate change induced inequalities, these are almost exclusively not the result of choices or faults of those individuals between whom it obtains, but the result of choices and faults of members of past generations.

[27] A possible exception here is Eric Rakowski (E. Rakowski, *Equal Justice* (Oxford: Clarendon Press, 1991), pp. 79–81). However, he stresses that exposure to climate risks has an element of option luck, whenever people could have chosen to live elsewhere and under more favourable climatic conditions.

[28] See also Page, 'Intergenerational justice of what', p. 461; Gosseries, 'Cosmopolitan luck egalitarianism and climate change', p. 288.

[29] G. A. Cohen, 'On the currency of egalitarian justice', *Ethics*, 99 (1989), 906–44 and K. Lippert-Rasmussen, *Luck Egalitarianism* (London: Bloomsbury 2015), ch. 4.

to wit, resources that increase the number of years that we can live healthy lives. Few would deny that a reduction of this number will reduce our well-being. They should find it hard to explain why increasing length of life does not increase our well-being. Also, the present line of argument suffers from having to appeal to the puzzling claim that the climate, unlike all other resources, affects our well-being.

So far I have argued that, from the point of view of luck egalitarianism, it is better if, in the reduced growth scenario, the present generation pursues the 'Don't care for the future' policy. If instead the present generation pursues the 'Reduce emissions' policy, the future generation will be even better off than the two previous ones without this in any way reflecting differential exercise of responsibility. Some might want to strengthen this conclusion. They might say that even if reducing our level of GHG emissions would not harm us in any way, but would benefit the future and better off generations, it would be in one way bad if we reduced this level. It would be bad, from a luck egalitarian point of view, even though it harms no one in that it would increase intergenerational inequality. This striking implication, which is well explored under the name of the levelling down objection,[30] some might say, is very implausible and shows that in relation to climate justice (also), luck egalitarianism has some very implausible implications.[31]

I have two brief responses to this objection. First, I am not sure that the levelling down objection represents a knock-down argument against egalitarianism. As Temkin has argued many people subscribe to other values, e.g., desert, that also imply that there could be scenarios where some are worse off and no one is better off, and yet this scenario is in one respect better.[32] Also, in response to the levelling down objection many endorse prioritarianism as a way of accommodating their egalitarian intuitions. Yet, it is unclear if prioritarianism is not vulnerable to a similar objection.[33] These responses are well-rehearsed elsewhere and have nothing to do specifically with climate chance and I can only sketch them here.

There is a more specific reason why the levelling down objection may not apply in this particular case and this is my second reply. Suppose we, i.e., members of the present generation, choose the 'Reduce emissions' policy and thus increase intergenerational inequality by reducing our GHG emissions (which, ex hypothesis, leaves us neither worse, nor better, off) and as a result

---

[30] Parfit, 'Equality and priority'; L. S. Temkin, *Inequality* (Oxford University Press, 1993), pp. 245–82; N. Holtug, *Persons, Interests, and Justice* (Oxford University Press, 2010), pp. 169–201.
[31] E. Page, 'Distributing the burdens of climate change', *Environmental Politics*, 17 (2008), 556–75, p. 564.
[32] Temkin, *Inequality*, pp. 245–82.
[33] I. Persson, 'Why leveling down could be worse for prioritarianism than for egalitarianism', *Ethical Theory and Moral Practice*, 11 (2008), 295–303.

make future better off generations even better off thereby increasing intergenerational inequality. The resulting additional inequality is then a result of a choice that the present generation has made. Hence, as far as this extra inequality is concerned, the present generation cannot say that it is worse off through no choice of its own, so according to the formulation of luck egalitarianism given above this additional inequality is not unjust. Accordingly, luck egalitarians might say that this particular case of levelling down is not better even in respect of equality. If the worse off present generation chooses a climate policy that leaves the future generation better off, while not affecting the level of well-being of the present generation, then the additional inequality is a result of a choice made by the worse off and for that reason it is not bad, not even regarding equality.[34] This scenario may not be particularly plausible – for instance, it may not be particularly plausible that the worst off members of the present generation would agree to make themselves worse off for the sake of much better off future generations. However, even if this is so, it would not show that the 'Reduce emissions' policy could not be implemented in a way that it would not result in an increase in unjust intergenerational inequality (despite the fact that it would result in more intergenerational inequality, descriptively speaking.)

It might be replied that the additional inequality that results from the present generation's abstaining from levelling down involves that future generations are even better off through no choice or fault of theirs and, surely, it is unjust from a luck egalitarian point of view if some people are *better off* than others through no choice or fault of theirs.

This reply can be dismissed, however. It is open for luck egalitarians to argue that while, as a matter of conceptual necessity, one generation is worse off than another if, and only if, that generation is better off than the former, what makes the distribution unjust is that one generation is *worse off* through no choice or fault of their own and not that one generation is *better off* through no choice or fault of their own.[35] Since in the abstaining-from-levelling-down scenario I imagine the present generation is not worse off to an additional degree through no choice or fault of their own there is nothing regrettable from the point of view of equality if the present generation acts in a way that leads to its being even worse off than future generations.

---

[34] I would want to qualify this conclusion in cases where present generations reduce their emissions, because they believe they are morally required to do so (K. Lippert-Rasmussen, 'Luck-egalitarianism: faults and collective choice', *Economics and Philosophy*, 27 (2011), 151–73). As far as such cases are concerned, I would then have to rely on the first reply to the leveling down objection only. Also, the present reply ignores the increased inequality between *past* generations and *future* generations benefiting from the *present* generation's reduction of GHG emissions.

[35] K. Lippert-Rasmussen, 'Hurley on egalitarianism and the luck-neutralizing aim', *Politics, Philosophy, and Economics*, 4 (2005), 249–65.

Luck egalitarian injustice and inequality-generating global warming

There are, however, different developments resulting from global warming other than the one represented in the reduced growth scenario that luck egalitarians may condemn. First, global warming may affect different people within the same generation differently. Specifically, it may increase the amount of inequality not reflecting choice within generations.[36] For instance, global warming may involve a much greater loss of gross domestic product (GDP) per capita in poor states than in rich states.[37] The latter group does not emit much GHG (but they may contribute to global warming in other ways, e.g., by having many children and by cutting down trees) and tend to end up worse off as a result of something the making of which they play only a very small part. Luck egalitarians might think that the most efficient way of addressing these inequalities is not by reducing GHG emissions, but by improving the conditions of poor farmers in other ways, e.g., by providing them with access to clean water. This may be true. However, at the level of policy it may be much less unrealistic to hope for a reduction of GHG emissions than to hope for significant direct redistribution of global wealth, because people in rich parts of the world see themselves as directly affected by global warming, but not by poverty in Africa.[38]

Second, the reduced growth scenario may not adequately represent the long-term future. It may be that after some generations of rising GHG levels global warming has disastrous effects on agriculture resulting in global famine or, as it is sometimes suggested, it may involve threshold effects such as stopping the Gulf Stream resulting in a new ice age (in Europe). In the disaster scenario, but not in the doomsday scenario, our accepting burdens to stabilise climate changes may result in less unjust inequality between generations than if we do not cut present emissions of GHG and, thus, cause future generations to suffer an new Ice Age or worse.

---

[36] Page, 'Intergenerational justice of what', p. 455. Page also writes: 'Since unchecked climate change will tend to exacerbate existing inequalities between countries, the egalitarian will expect the developed countries to fund generous policies of mitigation and adaptation ... because this is the most efficient way of achieving the desired outcome of greater equality' (Page, 'Distributing the burdens of climate change', p. 564). It is true that, from a luck egalitarian perspective, developed countries (or, more precisely, rich people most, but not all, of whom live in developed countries) should fund the policies of mitigation and adaptation adopted within a generation. However, for reasons of intergenerational justice it is not true that these policies will be generous in the reduced growth scenario.

[37] N. Stern, *Stern Review of the Economics of Climate Change*, (2006), http://webarchive.national archives.gov.uk/+/http://www.hm-treasury.gov.uk/sternreview_summary.htm, last accessed 10 October 2014). This is not to suggest that inequalities between states matter *per se*. However, they may be correlated with inequalities between individuals, and these inequalities matter *per se* from a luck egalitarian perspective.

[38] P. Singer, 'One atmosphere', in S. M Gardiner *et al.* (eds.), *Climate Ethics: Essential Readings* (Oxford University Press, 2010), pp. 181–99, p. 185.

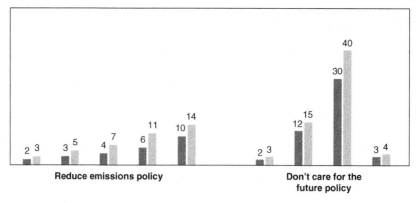

Figure 6.2

If Figure 6.2 represents the consequences of the two respective policies, luck egalitarians will prefer the 'Reduce emissions' policy. This seems plausible.

States and generations

Most of the inequalities resulting from global warming will obtain between members of different generations and between citizens of different states. Many, including some who have luck egalitarian sympathies, believe that such relations do not fall within the scope of principles of distributive justice.[39] Or that insofar as they do, these principles differ from the ones that regulate relations between citizens.[40] If either of these views is correct, luck egalitarians would have little reason to worry about global warming *qua* luck egalitarians.

The issue of the scope of principles of distributive justice is one that arises for any theory of distributive justice. Just as luck egalitarians may accept or reject a statist scope rejection on the principles of justice, so may friends of any other principle of distributive justice, e.g., sufficientarianism or prioritarianism. Hence, whether one should endorse such a scope restriction is an important issue, but it is not an issue that will give rise to any objections that specifically target luck egalitarianism.

In response to this claim, it might be said that if, among other things, luck egalitarianism is the view that makes it unjust for some to be worse off than others through no choice or fault of their own, and if people are members of their state or generation through no choice or fault of their own (as is often the

---

[39] T. Nagel, 'The problem of global justice', *Philosophy and Public Affairs*, 33 (2005), 113–47; J. Rawls, *The Law of Peoples* (Cambridge, MA: Harvard University Press, 1999).
[40] D. Miller, *National Responsibility and Global Justice* (Oxford University Press, 2007).

case), then luck egalitarians must reject the statist and generationist restriction on the scope of egalitarian justice. I agree, and this line of argument is similar to the one I pressed above in the section 'Does it matter if climate change is manmade?' in response to the view that if global warming is non-anthropogenic, it raises no worries regarding justice. Hence, luck egalitarians who think that only inequalities between co-citizens are unjust must give an alternative account of what makes these inequalities not reflecting differential exercise of responsibility unjust. Note, however, that something analogous can be said about competing principles of distributive justice. For instance, if what makes someone's level of goods unjust is that it is below the threshold of sufficiency, then a world in which all members of one state are above the threshold of sufficiency and all members of another state are all below the threshold of sufficiency is an unjust world. Some sufficientarians, of course, endorse the view that their favoured distributive principle has a global scope, but others may think that justice only requires that citizens make sure that no co-citizens fall below the threshold of sufficiency. These statists will face a challenge analogous to the one faced by statist luck egalitarians.

One influential defender of a statist view, David Miller, argues that luck egalitarianism fails to accommodate the value of national self-determination.[41] Assume for simplicity that nations and states are co-extensive. Miller's thought then is that global egalitarianism enjoins that people should be equally well-off whichever state they live in. Suppose we have two democratic states. One saves resources for the future – Ecologia – and the other – Affluenza – does not. As a result citizens of Affluenza end up being worse off over time. According to global egalitarianism as Miller construes it, this inequality is unjust and justice requires redistribution from Ecologia to Affluenza that displays collective irresponsibility in relation to its future. But this, Miller argues, fails to accommodate the value of national self-determination, because this value implies that a state can make collective political decisions without having to shoulder the costs of the collective decisions made by other states.

There are two reasons why this objection does not defeat luck egalitarianism. The first concerns how Miller construes global egalitarianism. As we have seen, luck egalitarians accommodate a concern for responsibility. Most contributions to the literature discuss the responsibility concern as a matter of individual responsibility. But this does not reflect any theoretical necessity. To the extent that there are collective choices that cannot be reduced to the set of choices of individual members of the relevant collective, luck egalitarians should incorporate this concern into their theory as well.[42] This might be done in different ways. The crucial point here is that just as standard luck

---

[41] Miller, *National Responsibility and Global Justice*, p. 73.
[42] cf. Gosseries, 'Cosmopolitan luck egalitarianism and climate change', p. 305.

egalitarianism does not condemn inequalities between individuals that reflect differential individual choices, collective choice luck egalitarianism does not condemn inequalities between collectives that reflect differential collective choices. Hence, while Miller has an important point, it is best construed not as an objection to luck egalitarianism, but as a point about how luck egalitarianism should be construed to accommodate the fact of collective choices.

My second reply is that specifically with regard to global warming the implication of global egalitarianism, as Miller construes it, does not seem implausible. If a state decides to reduce its level of GHG emissions and another equally well-off state decides not to do so, and if as a result of that difference in climate policies adopted, the latter state ends up better off than the former state, all other things being equal, redistribution in favour of the state that carried a greater portion of the common climate burden seems justified.[43] The state that decided not to do anything about climate change is simply free riding on the efforts undertaken by the other state and that is unfair. The important point is that, unlike in the kind of scenarios used by Miller to illustrate the clash between global egalitarianism and national self-determination, the effects of GHG emissions harm not just members of the state itself but everyone on Earth, and it is one thing to think that members of a state should be free to organise their own affairs and another thing to think that they should be free to harm members of other states. Hence, I conclude that not only is the problem of a statist or generationist restrictions on the scope of principles of justice not an issue that applies specifically to luck egalitarianism, but also that luck egalitarianism without such scope restrictions is plausible.

Responsibility

Luck egalitarians think that considerations of justice are crucially dependent on how individuals exercise their responsibility. In relation to climate this view motivates at least two interesting questions. First, for any given individual it is true that whether global warming takes place is something over which this person has no control.[44] However, because people might have options of

---

[43] One thing that would render all other things not being equal is that the transaction costs involved in redistribution are huge. In such a case redistribution from the climate irresponsible to the climate responsible state might not be justified all things considered, but it would still be fair and that suffices for the pro-global justice claim I defend here.

[44] D. Jamieson, 'Ethics, public policy, and global warming', in S.M Gardiner *et al.* (eds.), *Climate Ethics: Essential Readings* (Oxford University Press, 2010), pp. 77–98, p. 83. Strictly speaking this is false. Powerful individuals, e.g., US President Obama, might be able to affect global warming, albeit not through increasing, or decreasing, their personal emissions of GHGs.

moving, it is not true of any given individual that whether they are harmed by global warming is something which they have no control over. Still, global warming is a result of what many people do together and the option of moving might not be available to everyone. Indeed, some people take care not to emit too much GHG. The issue here is the extent to which negligible contributions to a bad, aggregate large-scale effect factor into luck egalitarianism or, for that matter, any other distributive principle that ascribes distributive significance to responsibility.

Second, of any given generation it is largely true that the climatic conditions which it faces are a result of non-human factors and what previous generations did. This would suggest that of any given generation it is true that the climatic conditions that it faces are not a matter of choices and faults by its members. Still, it may be true of a generation that it emits more GHG than any of its predecessor generations. Can it complain that it is bad for it to be worse off through no choice or fault of its own?

Note first that this question is different from the previous one. Here the focal issue is the fact that there is a considerable time-span between causes and effect. So this is an issue that would arise even if the last emission of each member of the present generation was decisive for whether the next generation would experience global warming. Hence, in principle at least one could say that tiny contributions are relevant to responsibility for being worse off, but that delayed effects are not, or vice versa.

Luck egalitarians might go two different ways here. They might insist that a generation that (i) is worse off than previous generations as a result of global warming and (ii) emits huge amounts of GHG that affects the climate under which future generations live is unjustly worse off, since its being worse off does not result from any choice or fault of its own or its members. Alternatively, luck egalitarians might revise the luck egalitarian formula presented in the introduction to this chapter and say that:

It is unjust for an individual to be worse off than others if, and only if, their relative positions (with regard to the relevant *equalisandum*) do not match their relative faultiness, i.e., how much at fault they are relative to others.[45]

On this view and assuming that faults are identified not just from the point of view of prudence, luck egalitarianism comes closer to some kind of comparative desert theory, i.e., one which says that it is unjust if people's relative positions do not reflect their relative deserts. On the assumption that faultiness is not defined only relative to prudence but also to morality, luck egalitarians

---

[45] Lippert-Rasmussen, 'Luck-egalitarianism: faults and collective choice'. Recall that I have resolved to ignore unjust equalities.

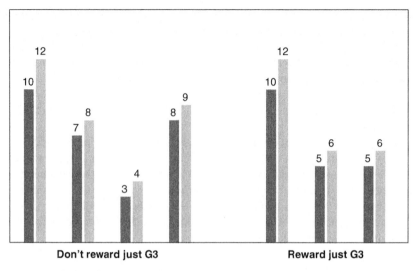

Figure 6.3

might say that a generation that emits high levels of GHG is at fault and for that reason it is not unjust if they are worse off than previous generations.

To see that this version of luck egalitarianism has plausible implications in this case, suppose that the first generation, generation$_1$, knows that generation$_2$ will emit large quantities of GHG thereby harming generation$_3$; that generation$_3$ will cut down on its GHG emissions thereby taking on a burden to benefit generation$_4$; suppose that, for some reason, generation$_1$ is better off than later generations; and suppose finally that generation$_1$ can either reduce its emissions of a certain GHG that affects the climate when generation$_2$ is around or reduce its emissions of a different GHG that affects the climate when generation$_3$ is around. These assumptions are represented in Figure 6.3.

On the first version of luck egalitarianism, generation$_1$ has no equality-based reason to choose the first scheme of GHG reduction, because generation$_3$ is worse off through its own choice to curb its GHG emissions, thereby benefiting generation$_4$, and had it chosen not to do so it would have been no worse off than generation$_2$, setting aside generation$_1$'s choice of reduction scheme. However, I think justice is best promoted if generation$_1$ reduces emissions of the latter GHG, thereby benefiting generation$_3$, and this supports my reformulation of luck egalitarianism.

Suppose the present generation makes itself worse off for the benefit of future generations. It is now even worse off than future generations as a result of its own choice to sacrifice resources for the benefit of the better off future generations. Does this mean that this extra degree to which the present generation is worse off than future generations is not bad from the point of view of equality? Again, cases such as these point to the need for revising the standard formulation of luck egalitarianism. In my view, we should understand responsibility for being worse off in such a way that individuals, say members of the present generation in the reduced growth scenario, who are worse off because they have acted in a way that they correctly believed themselves to be morally required to in order to benefit others, i.e., future generations, are not responsible in the injustice-defeating sense for their being worse off. Again this strikes me as a plausible implication.

Polluter pays principle

I have now completed my presentation of what luck egalitarianism implies with regard to climate justice and how it should be revised to accommodate various cases that are likely to arise in relation to global warming. I shall now compare luck egalitarianism so construed to some of the other principles of climate justice that have been proposed in the literature.[46] One such view is that those who have caused global warming should bear, or a suitable proportion reflecting their emissions of, the burdens of climate change.[47] This view is often summed up by the slogan 'polluter pays'.

Since the polluter pays principle enjoins distribution of climate burdens according to responsibility for global warming, and since luck egalitarianism similarly ascribes moral significance to facts about responsibility, initially one might think that the two principles are closely related. However, this is not the case. A polluter may be worse off than a non-polluter – e.g., the climate is harsher in the polluting country – in which case luck egalitarianism does not place the climate burden on the polluter unlike the polluter pays principle.[48]

---

[46] It might be said that the polluter pays principle does not have the same non-derivative status as luck egalitarianism. While this is true, a comparison of the two is not uninteresting. First, it is not as if there is no competition between luck egalitarianism and principles such as the polluter pays principle, i.e., they do have conflicting implications. Second, the polluter pays principle derives from a more fundamental principle, e.g., one that says that those who cause harm to others should compensate them, which does have the same non-derivative nature as luck egalitarianism.

[47] E. Neumayer, 'In defense of historical accountability for greenhouse gas emissions', *Ecological Economics*, 33 (2000), 185–92; Caney, 'Climate change, human rights, and moral thresholds', p. 136.

[48] This is not to deny that there are other cases where the recommendations of the two principles do not conflict. If, for instance, the polluting, worse off state is worse off through its own fault, e.g., it is at fault for its very inefficient use of energy which makes its citizens worse off, luck egalitarianism is compatible with the recommendation of the polluter pays principle, despite the fact that the polluting state is worse off.

In this case, luck egalitarianism strikes me as more plausible than the polluter pays principle.

There is another case where the two principles diverge and luck egalitarianism is more plausible than the polluter pays principle. Consider a case where the benefits of GHG emission accrue to individuals other than those who actually emit GHG. For example, suppose that certain industries that emit high levels of GHG are predominantly located in poor countries that export their produce to wealthy countries that, to their great benefit, as a result of this division of labour, can focus on knowledge-intensive, low-emission level industries. The polluter pays principle implies that in this case poor countries should shoulder the climate burdens. However, this view implausibly ignores who benefits from the relevant international division of labour.[49] Luck egalitarianism has more plausible implications in this case, although it does not focus on benefits tied to the emission of GHG (or, more precisely, to the activities involving emission of GHGs). There might be cases in which those who benefit from such activities are worse off than others. Here luck egalitarians will not want those who benefit to bear the climate burdens, since they are unjustly worse off despite these benefits. This again, however, does not strike me as an implausible implication.

There are other cases, however, where, arguably, the polluter pays principle is more plausible. Suppose a country has a choice between two different energy policies, which will leave it equally well off. The two policies differ, however, in terms of the amount of GHG emitted. If the country chooses the policy that involves higher emissions and thereby makes others worse off, it may seem plausible to require that it carries a greater burden of climate change than other countries even if this implies that it is worse off than other countries.[50] This goes against the canonical formulation of luck egalitarianism, but it is compatible with the revised principle that I have proposed in the section 'Responsibility'.

There is another dimension in which luck egalitarianism and the polluter pays principle seem to diverge, but where this reflects a particular narrow understanding of the polluter pays principle. This principle is often discussed on the implicit 'collectivist' understanding that 'polluter' ranges over nations and, thus, that present members of a nation are obliged to carry a climate burden reflecting the emissions of now deceased past members of the nation.[51]

---

[49] Gosseries, 'Cosmopolitan luck egalitarianism and climate change', p. 303.
[50] cf. K. Lippert-Rasmussen, 'Inequality, incentives, and the interpersonal test', *Ratio*, 21 (2008), 421–39.
[51] cf S. Caney, 'Cosmopolitan justice, responsibility, and global climate change', *Leiden Journal of International Law*, 18/4 (2005), 747–75.

This implies an obvious difference to luck egalitarianism, whose focus on choice and fault typically – but recall the previous section – is individualistic such that on typical luck egalitarian views members of the present generation of a nation cannot be held responsible for what past members of their nation did. However, there is nothing in the polluter pays principle that implies any particular view on how someone can be held accountable for what deceased persons, with whom they are somehow related, did. Hence, one version of the polluter principle says that only individual polluters pay and they pay only for their own pollution. On this view, there is no difference between the polluter pays principle and standard luck egalitarianism as far as the scope of individual responsibility is concerned.

Miller's principle of equal sacrifice

Partly in the light of some of the above-mentioned deficiencies with the polluter pays principle, David Miller has proposed a different principle for sharing the climate burdens which may seem egalitarian: the principle of equal sacrifice. According to this principle, 'targets for reducing gas emissions should be set in such a way that the costs of meeting these targets are allocated on an equal per capita basis among the members of better off societies. These costs will typically take the form of a reduction in projected living standards as a result of changes in methods of energy generation, personal lifestyle, and so forth.'[52] The measure of costs that Miller has in mind takes into account that the marginal reduction of well-being increases at lower levels of income. Accordingly, Miller envisages a 'more complex scheme of cost distribution where costs were zero for members of societies with endemic poverty, then rose in line with GDP until we reach middle-income societies, at which point they should be shared between all remaining societies on an equal per capita basis'.[53]

How does Miller's principle differ from luck egalitarianism? The most striking way in which it does so is that it, so to speak, assumes that the present unequal global distribution is the baseline distribution. Hence, assuming that Croatia is a middle-income and Norway a high-income country, Norwegians and Croatians should bear the same per capita costs on Miller's principle of equal sacrifice in terms of loss of projected well-being which would, according to Miller's more complex scheme, translate into the same monetary per capita costs. Luck egalitarians would reject this cost sharing scheme on the

---

[52] D. Miller, 'Global justice and climate change: how should responsibilities be distributed?', *The Tanner Lectures on Human Values*, Tsinghua University, Beijing, 24–5 March (2008), www.tannerlectures.utah.edu/lectures/documents/Miller_08.pdf, p. 146.

[53] *Ibid.*, pp. 147–8.

assumption that the inequality between Norway and Croatia does not reflect differential individual or collective responsibility. On their view, Norway should carry a much greater part of the climate burden. Which is the most plausible view in the light of this difference in implications?

Obviously, if money translates into power, then Miller's proposal has the distinct advantage that it is more likely to be the outcome of an international climate agreement than a luck egalitarian burden sharing agreement. However, this advantage is irrelevant to the question which I am asking here: namely, which principle is the more plausible principle considered as a principle of justice? Once we keep firmly in mind this is our question luck egalitarianism seems the better principle.

Suppose initially that the present level of emissions results from some countries aggressively pursuing economic growth through GHG emitting uses of oil and coal, and other countries restraining their pursuit of economic growth in the interest of not contributing to the greenhouse effect. Surely, if the former countries were to propose that the climate burdens were to be shared on the basis of Miller's equal sacrifice scheme, the latter might protest that costs are measured relative to present levels of economic wealth that reflect that some countries, but not others, have already restrained themselves in the interest of the global climate, and that it is unfair simply to disregard this, when distributing burdens in the future. In fact, these countries would seem to be able to object on exactly the same ground Miller gives for rejecting global egalitarianism in the previous section, i.e., that this principle clashes with the value of national self-determination because states are made to bear the costs of the collective political decisions of other states. In short, if Miller's proposal is to work we must be sure that the baseline relative to which we measure costs is justified.

In response, Miller might concede this point and then add that, as a matter of fact, our situation is very different from the one I imagined. Only quite recently was the correlation between global temperatures and GHG emissions established, so hardly any economic inequality between nations reflects that some have restrained economic growth in the interest of the global climate and others have not. Hence, the baseline is not unjustified for the reason stated in my objection even if, as time passes, it might be true in the future that it would be unfair for that very reason to take the existing economic levels of welfare as the baseline from which costs are measured.

At this point, luck egalitarians might press another objection, however. Suppose that the only food available is manna continuously falling from heaven. Some people are lucky to live where plenty of manna falls – at no time do they starve – and others are unfortunate to live where very little manna falls – they starve eleven weeks a year. For some reason moving people or manna around is impossible. Suppose that scientists discover that present levels of consumption of manna threaten the stability of the global climate such that to avoid climatic disaster the aggregate level of manna consumption must be reduced. Suppose finally that starving for a week reduces one's well-being equally much whether one starves regularly or never starves. In my view, people who starve regularly can object to a Millerean equal cost-sharing scheme on the ground that there is no justification for the unequal distribution of manna in the first place and, thus, no justification for considering it as the baseline from which costs should be measured. A luck egalitarian scheme where those who never starve should carry the climate burden would seem more just in this case.

Miller might concede this point, but then deny that global inequality is relevantly like the situation that I have depicted. The level of economic well-being enjoyed by nations, Miller might contend, to a very significant degree reflects different individual and collective choices made by their members and, accordingly, global inequality is to a significant degree justified and, thus, can be used as a cost-identifying baseline.

However, this reply has obvious drawbacks. First, to a very significant degree global inequality does not reflect differential individual and collective choice. This is certainly true of the inequality between the world's children who have neither made individual nor collective choices that would justify the tremendous inequalities that exist between them.

Second, luck egalitarianism can be formulated so as to accommodate facts about collective choice. Hence, even if Miller is right that such choices explain a significant part, or indeed all, of global inequality, this would not show that his equal sacrifice principle is superior to a luck egalitarian principle. It would simply show that, under this assumption (which I do not accept), to make sure that people's relative positions reflect nothing other than their relative exercises of responsibility, climate costs would have to be shared equally. I conclude that the principle of equal sacrifice is not superior to a luck egalitarian burden-sharing scheme.

Conclusion

In this chapter I have explored the implications of luck egalitarianism in relation to the distribution of climate benefits and burdens. I have argued that it has plausible implications. Also, I submitted that in many cases where some versions have controversial implications, similar implications can be identified in the case of competing distributive principles and, accordingly, these implications do not point to any flaws in luck egalitarianism *per se*. I also compared luck egalitarianism to two competing principles for how the climate burden should be shared: the polluter pays principle and Miller's principle of equal sacrifice. I showed how they differ from a luck egalitarian scheme and argued that neither is superior to such a scheme.[54] Finally, I have also conceded that, despite its popularity as a theory of distributive justice, luck egalitarianism does not identify the main reason why we have a moral obligation to reduce our emissions of GHG. Indeed, in the reduced growth scenario luck egalitarians would think that, regarding equality, it is in one way better if we do not reduce our GHG emissions!

[54] I thank an anonymous reviewer, Andreas Brøgger Albertsen, David Vestergaard Axelsen, Morten Brænder, Rasmus Sommer Hansen, Xavier Landes, Søren Flinch Midtgaard, Jeremy Moss, Lasse Nielsen and Tore Vincent Olsen for helpful comments on a previous version.

7 Individual duties of climate justice under non-ideal conditions

Kok-Chor Tan

On the institutional approach to justice, individuals' responsibilities of justice are defined institutionally and also limited institutionally.[1] That is, what particular duties of justice they have are given by the rules of just institutions, and their responsibility of justice is limited to complying with just institutional rules and to maintaining these institutions. But what responsibilities do individuals have in the absence of just arrangements? Is their duty to do that which counterfactually a just arrangement would require of them? Or does their responsibility include that of doing their part to bring about such an arrangement? I will discuss this question of personal responsibility in the absence of just institutions in the context of climate change justice.

Although I will clarify and highlight some of its basic assumptions and features, I largely assume the institutional approach for my purpose.[2] My aim is not to defend the institutional approach as such but to identify what its mandate, that individuals have a duty of justice to create just arrangements when these are absent, entails in the context of climate injustice. I will suggest that this special focus on institutions as opposed to taking direct personal action to curb emissions is not a misguided case of institutional-fixation.

I begin by taking the following for granted. At minimum, a collective response to global climate change must limit the rise in global temperature to 2°C, and meeting this goal requires reducing global carbon dioxide (CO_2) emissions due to human activities by 50 to 85 per cent of 2000 levels by 2050.[3]

---

[1] Many thanks to Jeremy Moss and a second reader for their very helpful questions and critical comments.
[2] The institutional view is most famously criticised by G. A Cohen, *If You're an Egalitarian, How Come You're So Rich?* (Cambridge, MA: Harvard University Press, 2000) and *Rescuing Justice and Equality* (Cambridge, MA: Harvard University Press, 2008); also L. Murphy, 'Institutions and the demands of justice', *Philosophy and Public Affairs*, 27/4 (1999), 251–91. I attempt a defence of it in K. C. Tan, 'Justice and personal pursuits', *The Journal of Philosophy*, 101/7 (2004), 331–62 and *Justice, Institutions and Luck* (Oxford University Press, 2012). See also D. Halliday's critical discussion in 'Review essay of *Justice, Institutions and Luck*', *Utilitas*, 25/1 (2013), 121–32.
[3] For example D. Moellendorf, 'Treaty norms and climate mitigation', *Ethics and International Affairs*, 23/3 (2009), 247–65, p. 249.

This raises the question of what each country's just emissions cap ought to be, and this question in turn raises the more fundamental question of global justice. As Henry Shue points out, there is a morally significant difference between emissions for development and subsistence purposes, and emissions for luxury purposes.[4] More comprehensively, from the perspective of global justice, a fair allocation of emission responsibilities among countries must account for their developmental needs, their relative abilities to take on the burden of emissions restrictions, their emissions history and their other just claims on each other.[5] Thus, for this reason, Moellendorf writes that 'any legitimate treaty [on emissions caps] must put much heavier mitigation burdens on industrialised countries'.[6] An account of just emissions taken to be part of the larger issue of global justice will allow for 'common but differentiated responsibilities', in which developed countries accept more demanding emissions responsibilities than developing ones.[7] The idea of differential responsibilities is not without its detractors, but this does not affect the structure of my present discussion. I note this only to provide one possible background scenario for just emissions to give some context to my discussion. The key point here is the need for countries to take on considerable (even if not differentiated) obligations to meet the global goal of slowing climate change.

Any emissions burden on a country will be a collective one that has to be distributed among its citizens and local associations. How this is to be fairly distributed among them is itself an important question of justice, in this case of domestic justice. We can safely assume that since the present per capita emissions of individuals in developed countries are well in excess of most reasonable accounts of globally just emissions per capita, individuals of developed countries will have to face certain downward adjustments to their current emissions expectations under any plausible just global emissions regime and however the allocation within a state is domestically assigned.

Thus any country, especially a developed one such as the US, Canada or Australia, that takes its just collective emissions cap seriously will need to establish a regulatory framework to restrict and control human CO_2-producing activities within its borders and to assign the burden of emissions restrictions among its constituents. This will entail regulating

---

[4] H. Shue, 'Subsistence emissions and luxury emissions', *Law and Policy* 15/1 (1993), 39–59.
[5] However, see S. Caney, 'Environmental degradation, reparations, and the moral significance of history', *The Journal of Social Philosophy*, 37/3 (2006), 464–82 for one discussion of the restricted relevance of history for climate justice.
[6] Moellendorf, 'Treaty norms and climate mitigation', p. 248.
[7] S. Caney, 'Cosmopolitan justice, responsibility, and global climate change', *Leiden Journal of International Law*, 18/4 (2005), 747–75, p. 772.

industrial and agricultural activities, imposing emissions standards on cars and appliances, enacting a carbon tax on commercial and personal consumption, re-conceiving the way public and shared commercial spaces are used and maintained, introducing new forms of urban design and housing, providing better public transportation, providing governmental support for developing clean energy alternatives, imposing legal restrictions on drilling and fracking and so on. Much of the effort will be directed at commercial and industrial activities as well as at infrastructural improvements (such as in energy production and regulation of industries). No doubt these background and infrastructural reforms will impose certain burdens on individuals, who will have to pay more for their personal energy consumption, the goods they buy and the like. But restrictions directly targeted at personal behaviour will also be required, including, for example, introducing limits on how much waste individual households can produce each week for collection, imposing higher gasoline taxes and limiting usage of private vehicles in urban centres and so on. So under any regulatory arrangement that is compatible with the just emissions cap of a country, individuals can be legitimately expected to accept certain burdens and to make some adjustments to their emissions and emissions-related expectations.

On the institutional view of justice, individuals have the duty of justice to comply with the demands of just institutional arrangements. Their personal just entitlements will be set by the parameters of the just institutional order of their society, and any restrictions on their lifestyle choices, including caps on their emissions entitlements that a just climate regime would imposed on them are legitimate, that is just, restrictions. Their duty of climate justice then will be to comply with and support the rules of the just regulatory framework of their society. This climatic duty, even though institutionally focused, is not to be mistakenly interpreted as an overly lax duty. Compliance with the rules of an ideally just regime, as noted above, will require important adjustments in personal emissions expectations and will constrain the range of available personal choices and options in consumption, habits and expectations. Moreover, the maintenance of a just regulatory framework will require a certain attentiveness, awareness and vigilance on the part of individuals. They have to acknowledge the demands of climate justice and be reasonably informed as to whether their institutions are in fact equal to these demands. This requirement of *institutional vigilance*, as I might call it, follows from the duty to support and maintain just institutions, for without some reasonable requirement that individuals be attuned to the operations of their institutions, and be equipped to respond should their shared arrangements tend towards injustice, the duty to support and maintain just institutions rings hollow. Vigilance will also require individuals to be adequately politically engaged, and to be ready to keep their legislators in line should they make institutional decisions at odds with what justice demands. The requirement of vigilance does not entail that each has to strive for complete epistemic certainty which, if it is not an impossibility, then objectionably overdemanding of individuals' agents. What is required, rather, is that they strive for a certain literacy with respect to justice as is appropriate to their social and cultural circumstance. The requirement of vigilance does not fault the average individual for failing to know that their institutions were inadequate to the demands of climate justice thirty years ago. But it is not unreasonable to say that vigilance would require that a person in a developed society today not be wholly ignorant at least of the concerns and issues surrounding climate change and its proposed causes. These points are, I hope, relatively uncontroversial. I recount them only to emphasise the fact that the duty to comply with and to maintain just institutions does not let individuals off the hook easily. An institutional focus for justice does not result in an under-demanding account of justice.

But what responsibility of justice does a person have in the absence of a suitably just regulatory scheme? The parameters of individual responsibility provided by the institutional approach are not available under this non-ideal condition. Yet this is the typical situation with respect to climate justice. Societies like the US, many would argue, lack an emissions regime that satisfies its global climatic obligation. What duties of climate justice do individuals in such societies have in the absence of just institutions of this kind? To focus the discussion more vividly, let us assume that we are concerned with persons who accept that the existing regulatory regime in their society is unjust and that climate justice requires a more demanding emissions-restriction arrangement. Thus our question can be framed as follows: If you believe that your society is not doing its part (collectively) to mitigate climate change because it lacks just institutional rules, what responsibility of climate justice do you as an individual have?

My concern in this chapter is with how an *institutionalist* with respect to justice can respond to this question. An institutionalist, as I understand it, is one who holds that justice is concerned primarily with the basic institutions of society and that the rules of just institutions define the content of persons' duties of justice and other obligations.[8] This approach to justice is famously

[8] The institutional view pertains to the site of justice, and this does not as such rule out ideals of international justice (such as a cosmopolitan one). Institutional justice can also be global in scope if the relevant institutions to which justice applies are global in reach. If international justice, as it will on the institutional view, means that there must be international institutions of an appropriate kind, then it will be the responsibility of individuals directly but also more often through their respective state institutions to affect and maintain these international institutions. With respect to global regulatory frameworks on emissions regulation, it is typically states that must assent to these arrangements (rather than individuals directly). In this case, a state's failure to support a just global climate regime is a mark against the justness of the institutions of that state (for example, in failing to enact laws that will ratify global climatic agreements). But on the institutional approach, citizens of that state will have the duty to remedy this institutional failure. I discuss the institutional *site* and the global *reach* of distributive justice in Tan, *Justice, Institutions and Luck*.

represented by John Rawls, who writes that 'within the framework of background justice set up by the basic structure, individuals and associations may do as they wish insofar as the rules of institutions permit'.[9]

Since duties of justice are defined institutionally on this approach, the absence of just institutions thus presents a problem for the institutionalist that a non-institutionalist avoids. For example, a utilitarian who holds that it is the end result that matters and that therefore institutions are not morally significant in themselves, will not see the absence of just institutions as a special problem. The absence of institutions might present an instrumental problem – i.e., the absence of a means of affecting a desired outcome – but it does not pose a distinctive problem of justice on its own. In this discussion, I presume here the institutional view. My aim in this chapter is to see what institutionalism can say about the problem of climate justice in the non-ideal case where just climatic institutions are absent.

Different options in the absence of just institutions

As mentioned, when just institutions are absent, there are no just institutional rules to define the duties of justice for individuals and to provide the parameters of admissible personal pursuits. What responsibilities of justice do persons have in this case? What responsibilities of justice do persons have if they know that their current emissions regulatory arrangement is not sufficiently just and that a more just one would require further constrains on their personal pursuits?

To better clarify what the institutional approach, as I understand it, enjoins in this case, let us consider four general options in response to this challenge. The aim here will not be that of providing decisive arguments against the alternative options but more modestly to highlight the different assumptions motivating the institutional option, and to suggest why an institutional focus is not implausible as a response to climatic injustice.

(a) All bets are off: since justice is given institutionally, and since there are no institutions to define justice, nothing can be just or unjust. So individuals are not required to do anything as a matter of climatic justice in the absence of any climatic arrangements or institutional requirements. If there are some institutional demands, even if these demands are hardly adequate to the cause of climate justice, individual responsibility is to do just that that is defined institutionally. Thus this extreme form of institutionalism takes it rather literally (and with apologies to Hamlet) that there is nothing just or unjust but institutions that make it so, that there is no independent expectation of justice other than what is institutionally laid out.

---

[9] J. Rawls, *Justice as Fairness* (Cambridge, MA: Harvard University Press, 2001), p. 50. See also J. Rawls, *A Theory of Justice* (Cambridge, MA: Harvard University Press, 1971), pp. 7, 115 and 110.

(b) Do all you personally can to promote justice: in the absence of a just institutional arrangement, individuals must take it upon themselves to do what they can, as much as they can, to realise the ends of justice. That is, individuals have to take on the burdens of climatic justice responsibilities on their own, and if the absence of a just arrangement means that many others are not doing their part, individuals have the responsibility of justice to pick up this slack when they can.

(c) Do (only) what just institutions would require of you: in the absence of a just arrangement, individuals are to identify what a just arrangement would require of them, and their duty of justice is to do that which would counterfactually be required of them. Their duty is thus limited to what institutional justice, counterfactually, would demand of them were it realised. Their duty does not include taking on any extra that is due to others not doing their part.

(d) Help create just arrangements: the most important responsibility of justice of individuals is to do their share to create or bring about just arrangements.

Option (a), *All bets are off*, is in a sense a kind of institutional view. But it is tied to, what I will loosely call the Hobbesian idea, that until there are institutions there can be no expectations of justice. The institutional view I am presuming, on the other hand, is more Kantian than Hobbesian (to provide a countervailing loose historical contrast). The Kantian account holds that while duties of justice are given institutionally, there is a pre-institutional responsibility of justice to create and maintain an institutional order if this is what is necessary for the realisation of a just social order (an order in which the rightful claims of all can be respected).[10] That is, even though there are no just arrangements specifying individual duties of climatic justice, it does not follow that individuals do not have the shared responsibility of justice to bring about a just arrangement. This view is well illustrated in Rawls's theory of justice by his remarks that we have the natural duty of justice to create just arrangements when they don't exist.[11] The institutional view I am adopting holds that justice's demands are specified institutionally, but there is the pre-institutional duty of justice to bring about the appropriate institutions. The Hobbesian institutional view accepts the first part of this thesis, but denies the second. So this option is not on all fours with the institutional approach I am assuming.

---

[10] Kant writes: 'When you cannot avoid living side by side with all others, you ought to leave the state of nature and proceed with them into a rightful condition, that is a condition of distributive justice [that is possible only under a public system of laws]'. I. Kant, *The Metaphysics of Morals*, M. Gregor (trans.), (Cambridge University Press, 1991 [1797]), p. 122. Compare with Hobbes who says: 'The notions of right and wrong, justice and injustice have there [in the state of nature] no place'. T. Hobbes, *The Leviathan*, J. C. A. Gaskin (ed.), (Oxford University Press, 1996 [1651]) p. 85. This is not to say that for Hobbesian, we have no reason to create institutions that will then define our rights and duties of justice. We have reasons of rational interests to leave the state of nature but not reasons of right or justice.

[11] Rawls, *A Theory of Justice*, p. 115.

Of course the fact alone that option (a) is not in line with the institutional view does not show it to be false. My main purpose in identifying this option is to provide a foil against which to further clarify the institutional approach I am presuming. Still, it is worth noting a difficulty with this option. It takes it that absent institutions there is no personal duty of justice at all. This will not hold sway over us if we think that the requirement of climatic justice is not conceptually tied to institutions in this way, such that the absence of just climatic regulatory institutions means that people have no duties of climatic justice at all.

Option (b), *Do all you personally can*, will appeal to some non-institutionalists, such as the utilitarian. Institutions on this approach are instrumentally valuable only; they are useful solely for helping bring about a particular state of affairs. Thus where institutions are absent, individuals can be expected to do whatever it takes to realise the end that it is the function of institutions to promote.

But this option is plainly at odds with the institutional approach as I will recount. First, the institutional approach presumes a kind of value pluralism. An underlying motivation for the institutional approach is the presumption of the pluralism of the good, that is, that there are different valuable conceptions of what the goods and ends in life are. The aim of an account of justice is to establish the parameters within which persons may form and pursue their ends in life on fair terms with each other. But while justice sets the fair terms and by doing so provides a framework that makes legitimate personal pursuits possible, it is also not the case that the pursuit of justice is the definitive goal of any individual's life, such that a life that is not primarily devoted to the cause of justice and defined by this pursuit is a failed personal life. An individual's life fails to be a good one if they realise their ends unjustly; but their life is not a failure just because it is not one wholly devoted to the end of justice. We might say that justice has regulative primacy over other ends and goods, but justice is itself not a dominant end in the sense of it defining or serving as the goal of all valuable human engagements. The institutional approach, as I understand it, begins from this value pluralism. It takes it that an institutional focus for justice provides a way of interpreting and thus identifying the site of justice so as to secure its primacy without denying the importance of other values.[12]

The directive that persons have to do what they can to bring about a just end takes justice to be a dominant value is opposed by value pluralism. It interprets the demands of justice too broadly and draws the limits of personal life too narrowly.[13] To insist that all personal efforts serve to realise a just state of affairs defies a basic motivating assumption of the institutional approach,

---

[12] This is the subject of Part I of Tan, *Justice, Institutions and Luck.*
[13] cf. J. Rawls, *Political Liberalism* (New York: Columbia University Press, 1993), p. 174.

which is that we need to interpret the demands of justice and the demands of personal life so that these two values hang together acceptably. An account of justice that does not allow sufficient space for personal pursuits, but which allows for personal pursuits only if they are in the service of justice, will be an interpretation of justice that fails to acknowledge the complexity of moral ends people have. On the other hand, an understanding of personal life that can *a priori* limit the requirements of justice runs against most reasonable understanding of role of any conception of justice, which is to set limits on personal claims.

The institutional approach to justice allows an interpretation of both justice and personal life that allows space for valuable personal pursuits while preserving the regulative primacy of justice.[14] By training justice to the design of institutions, just background conditions for personal interactions are secured; but by limiting justice's demands to institutions, personal space within the rules of institutions is preserved.

Stating the motivating assumption behind the institutional approach of course does not amount to a defence of it. But it helps us appreciate its appeal more by clarifying the problem that it is designed to address. Of course there is the question of how extensive the reach of institutions ought to be, and different institutionalists will have different views on this substantive matter. For example, how extensively should the state regulate, say, matters like the raising of children? Or how much state involvement should there be with regard to spousal relationships within the family? These are questions of what should fall within regulatory institutional ambit of the state, of what should fall within what Rawls calls 'the basic structure' of society. Responding to these substantive questions will invite considerations of not only what the state can in practice regulate, but also what it can in principle or permissibly regulate. But the point common to all institutionalists is the formal or structural point: that there is a moral division between the demands of institutions and demands for personal life.

It is also worth noting that option (b) does not adequately reflect the collective character of justice. As a collective project, justice is not fully secured by a person doing all they can; it requires that society as a whole adopt a certain ideal reflected in its social arrangements. Option (b), to be sure, can require that individuals, for the better discharge of their individual duties, collaborate and coordinate their actions. But the collective aspect of justice I am suggesting that this option bypasses is not merely that we need to coordinate individual efforts. Rather it is that what justice requires is more than the sum of individual actions. For example, each of us doing all we personally reasonably can while keeping institutional structures constant, will not reduce our society's

---

[14] Tan, *Justice, Institutions and Luck*, pp. 26–32.

Individual duties of climate justice 137

level of emissions climate significantly. What would be required, in additional to personal effort, are changes in our industrial and agricultural rules and practices – a major source of emissions. This collective character of justice is a point I will return to later.

To sum up the discussion so far: option (a) offers an overly contingent account of justice (to wit, taking the actual presence of institutions to be one of the preconditions of justice) and option (b) offers an overly stringent interpretation of it by treating it as an all encompassing and pervasive moral pursuit. Option (c) is more compatible with the institutional approach and I will turn to it next.

Do what just institutions would require

I take option (c), that individuals are to *do what just institutions would require* of them, to line up with the institutional approach in a way that option (b) does not. Under option (c), the scope of individual duty is given and limited institutionally – one has to reduce emissions up to what a just institutional arrangement, were it in place, would require of one. But it is still not an institutional duty in the sense that even though the scope of the duty is defined institutionally, it is not a duty directed at institutions in that it is not aimed at correcting or reforming institutions. We can call it a *non-institutional duty* to contrast it with the duty to create institutions. Examples of non-institutional duties in the context of climate justice include reducing household emissions by improving insulation and cutting down on energy consumption at home, installing energy efficient appliances, driving less, reducing consumption, recycling, cutting back on non-essential travels and so on. Under option (c), one is to reduce one's overall emissions through these personal actions up to the point that would be expected of one under a just regulatory arrangement.

It will be helpful here to consider the 'compliance condition' on the principle of beneficence as proposed by Liam Murphy. Murphy's compliance condition holds that our obligations of beneficence ought not to demand more of agents than would be demanded of them under full compliance.[15] That is, our duty does not increase just because others are doing less. To support the compliance condition, Murphy argues, beneficence is best understood as a collective and cooperative project rather than an individualistic one. As a collective project, the burdens of beneficence have to be assumed by all members of the collective. It undermines the collective character of this burden if one's share of duty increases just because others fail to do their part.

---

[15] L. Murphy, 'The demands of beneficence', *Philosophy and Public Affairs*, 22/4 (1993), 267–92, p. 277.

Now Murphy's compliance condition applies to the duty of beneficence rather than justice. In fact, Murphy elsewhere rejects the distinctiveness of a duty of justice as understood on the institutional view.[16] For my purpose, I adopt the structure of Murphy's account without endorsing his substantive commitments. Applying Murphy's condition to personal duties of justice, the proposal then is that one's duty of justice under conditions of non-compliance does not exceed one's duty of justice as defined under full compliance. Thus in the absence of a just regulatory framework, one's duty of climatic justice does not exceed that which would have been defined by a just regulatory framework.[17]

Unlike option (b), which takes our duty of justice to be individualistic, the present option duly regards justice as a collective project and thus sees that it makes a difference whether 'an increased need is due to non-compliance by others'.[18] As said, understood as a collective project for which all members are to do their part, my burden of justice ought not to increase merely because others have neglected their duty. Perhaps there are other possible explanations for this compliance restriction on person's duties; perhaps there ought to be some other condition of limitation. At any rate, option (c) with some limits on what persons can be expected to do in response to climate justice seems independently more plausible than option (b). Unlike (b), (c) is not objectionably overly demanding.

Thus, on this approach, while institutions serve an important framing function for personal responsibility, there is no requirement that individuals go about realising just institutions. It is sufficient that they imagine what just institutions would be like and what these institutions would counterfactually demand of them. They are then to do their part as these institutions would dictate should they exist.

Yet there is a limit to applying the compliance condition to the problem of climate change justice. What would a just emissions regulatory framework require of each of us personally? How much we can legitimately consume and how much we can each produce, how much we should each not drive, how regularly we travel, depend on background factors such as how much energy is used in producing the things we consume, how waste is handled and disposed of in our society, how fuel-efficient our cars are and so on. Our personal obligation with respect to climate change is rather under-defined without

---

[16] Murphy, 'Institutions and the demands of justice'.

[17] In short, I take Murphy's compliance principle to be separable from his anti-institutionalism, and attachable to an institutional approach in the following way: in the case of unjust institutions, persons are to do as just institutions would require of them (as opposed to Murphy's own view which is that persons are not required to do more than what their share would be under full compliance, when everyone is doing their part).

[18] Murphy, 'The demands of beneficence', p. 286.

some just background regulatory structure. In the absence of this structure, our duties necessarily remain indeterminate since it is the structure that defines our duties.

By indeterminate, I mean that the problem is not just an epistemic one (although there is also this problem) of persons not knowing what their duties are even though there is a fact of the matter. Rather it is that their duties can be under-specified and undetermined without institutions to define them. For example, we can each know that we need to reduce CO_2 emissions. But what does that require of us? How much are we to cut back on? What kinds of 'restitution' in terms of carbon tax rates, for example, should we each be required to accept? There is no pre-institutional determination of these matters. There is no pre-institutional fact of the matter of what our specific duties of justice are.[19]

In other words, how the collective burden of emissions restriction is to be exactly assigned among individuals of a society is undefined in the absence of institutions to define each person's share of the burden. One reason for this is that there are different possible domestic allocative arrangements that would satisfy the requirements of justice. To illustrate this last, imagine for simplicity that climatic justice in a society will require some combination of personal/household carbon tax, restrictions on household waste and adjustments to the price of gasoline. Climate justice under this simplified scenario will require some coordinated carbon tax scheme, waste-restriction and gasoline tax. What an individual should pay by way of their carbon tax will depend on how the other two variables are regulated. Different combinations of taxes, waste restrictions and gasoline consumption control are available, and which particular *institutional set* one ought to conduct oneself by is a matter for society to collectively decide. The matter is further complicated by the fact that the burdens will be differentially assigned among individuals of the society depending on their income, ability and so on. Absent a publicly known and endorsed institutional arrangement, individuals cannot on their own know what their institutional duty of justice is because this duty remains indeterminate.

Can this problem of indeterminacy be evaded by appealing to some ideal of just per capita emissions for a country (as determined by some global convention), and take that per capita ideal to be the definitive target for individuals of that country? So, even if a state fails to implement just regulatory arrangements domestically that can specify individuals' obligations of climate justice, if there is a global standard that can identify that society's just per capita emissions, do individuals not know what their obligations of climate justice are in spite of

---

[19] This again, to recall the Kantian point, explains why we have the responsibility of justice to create institutions (to leave the state of nature) in order that our duties of justice can become more determined.

domestic institutional failures? The reason, however, why this move does not work is the obvious fact that any computation of per capita emissions for a country includes 'direct' and 'indirect' emissions. Indirect emissions are emissions due to industrial, agricultural and other corporate and collective activities. Without institutional specification in the domestic domain to allocate and allocate these indirect collective emissions among citizens of a country, the indeterminacy problem remains even if there is some agreed on global per capita level of just emissions to aim for.

The above points to an implication that is worth mentioning even if obvious because it will reiterate the collective character of the problem of climatic justice. Using some ideal of just per capita emissions as a guideline for *personal conduct* applies an evaluative standard to the wrong thing. It uses a measure for collective conduct as a guide for personal behaviour. Indirect emissions make up a large portion of per capita emissions, often in excess to what each person emits directly through their personal activities and consumption choices. Thus if some ideal of just per capita emissions is invoked to set the standard for personal conduct, it will not demand very much of them personally. The requirement that we each constrain our direct personal emissions against some just per capita ideal will be overly lax since much of per capita emissions include collectively produced emissions outside the scope of personal choice. The point here is not offered against option (c)'s claim that when persons do their share as specified by that ideal, we realise justice. In fact (c) does not say this. My objection here is that if this is what the option requires, it will demand require very little by way of individuals with respect to climate change. Most ideals of per capita emissions quota far exceed the levels individuals in their personal life actually produce.

To illustrate: according to the US Energy Information Administration (EIA) per household emissions in the US were 59 tonnes of CO_2 in 2003. Yet of the total 59 tonnes, only 24.1 tonnes come from 'direct emissions' – 12.4 for household operations and 11.7 for gasoline. The remaining approximately 35 tonnes emitted that is attributed to each household comes from indirect emissions, such as commercial, industrial and transportation activities.[20] Suppose we determine, not conservatively, that a just emissions cap would reduce per household emissions to 30 tonnes (thereby significantly reducing the level of per household emissions by roughly half). If we apply this standard under the compliance condition to determine our *personal household* responsibility, then our household obligation can be literally discharged without any need to adjust personal direct emissions since that already comes in well under the level of 30 tonnes, at 24.1 tonnes. So it is not only that this standard is under-demanding – it in fact

---

[20] EIA numbers obtained from Hinkle Charitable Foundation, *How Do We Contribute Individually to Global Warming?*, www.thehcf.org/emaila5.html, last accessed 27 October 2014.

demands nothing at all from individuals. The problem here is that as stated previously: transferring to individuals a standard of responsibility that is meant to address a collective problem.

Now one might ask: why not expect each of us to do all we can to reduce actual per capita emissions, including making up for indirect emissions, to realise the ideal just per capita emissions? But this move takes us back to option (b): if we are each to do all we personally can to reduce per capita emissions, given that much of this is due to indirect emissions that even extraordinary personal sacrifices would not affect very much in the absence of societal reforms, this will be an unreasonably overly demanding requirement, not to mention futile.[21] One may retreat from the world, live the life of an ascetic, reducing one's personal net emission to nearly zero, and this self-sacrifice will still not make a difference to climate change if society continues using energy as usual and social institutions remain the same.

In short, option (c) is either under-determined or insufficiently demanding. It is insufficiently demanding if it gets its standard for personal responsibility from some global ideal of just per capita emissions. Since that ideal is a specification of a society's collective duty, it will not demand much, if at all, if proposed as a guideline for personal conduct. It is undemanding, again, in that it will literally ask nothing of individuals at all. On the other hand, if we want something more demanding, that will compel some personal response, we will need to imagine what just institutions would require of us. But in the absence of actual institutions, we face the problem of indeterminacy.

The question we are addressing is what climate justice demands of persons in the absence of a just regulatory framework. My point is not that individuals are under no obligation to reduce personal emissions where they can. While their duty is under-determined, it is still possible that each is required to reduce their emissions quite considerably. I am saying here that justice has to require something more in addition to that. Whatever the merits of personal reduction of emissions (and there is much to commend here), climate justice I will suggest will require the establishment of a just regulatory framework. That is, it will require that we each do our part to create the right institutions.[22]

---

[21] For example, L. Meyer and P. Sanklecha, 'Individual expectation and climate justice', *Analyze and Kritik*, 2 (2011), 449–71, pp. 462–63; P. Singer, *Practical Ethics*, 3rd edition (Cambridge University Press, 2011), p. 231, 235.

[22] In contrast, see Sinnott-Armstrong who argues that there is no personal moral obligation with respect to climate justice (his example is that of avoiding wasteful driving) but the obligation to help bring about better social institutions: W. Sinnott-Armstrong, '"It's not *my* fault": global warming and individual moral obligations', in W. Sinnott-Armstrong and R. B. Howarth (eds.), *Perspectives on Climate Change: Science, Economics, Politics, Ethics*. Advances in the Economics of Environmental Resources (Amsterdam: Elsevier, 2005), vol. V, pp. 285–307.

The duty to create institutions

Let us turn then to option (d), *Help create just institutions*. Recall here Rawls's oft cited statement of the natural duty of justice: 'the most important natural duty is to support and to further just institutions. This duty has two parts: first we are to comply with and to do our share in just institutions when they exist and apply to us; and *second we are to assist in the establishment of just arrangements when they do not exist,* at least when this can be done with little cost to ourselves'.[23]

It is the second part of the natural duty, the italicised part, that concerns us. This natural duty to establish just arrangements, which I will call an institutional duty, can be contrasted with an interactional duty such as the duty to provide assistance directly to another.

This institutional duty can be interpreted in three ways. It can be given an *exclusive* reading. On the *exclusive reading*, the duty to help create just institutions is what justice requires *and* it is all that justice requires. That is, a person does not discharge their duty of justice, in the absence of just institutional arrangements, by taking on direct personal action like providing direct assistance to others. In the case of climate justice, one does not fulfil one's obligation of justice by simply reducing personal emissions (by cutting down driving, household consumption, etc.) even if one accepts that one has a moral duty to do these personal acts under unjust conditions. One's duty of justice includes necessarily doing one's part to help bring about just arrangements. Personal actions may bring some good, but they do not qualify as acts in the service of justice on the institutional reading.

But while justice requires persons to act on their institutional duty, justice does not require them to do more. Justice covers and only covers one's institutional duty. An agent may have other non-institutional and personal duties, but these are not duties of justice but other kinds of moral duties like duties of beneficence, mutual aid and so on.

The institutional duty can be given an even more restrictive reading, which I will call the *extreme version*: on this reading, not only is the institutional duty necessary and sufficient for justice, but it has general priority over all other moral duties. When there is a trade-off between acting to promote good institutions and acting directly on beneficence, the former takes precedence. So the extreme reading adds the condition of priority to the strong reading.

Finally, there is the *inclusive reading* of the natural duty: justice requires that a person does their share to bring about just arrangements, but it is possible that in addition to this institutional duty, an agent can have the duty of justice to take personal action as well. So the inclusive reading accepts that the institutional

---

[23] Rawls, *A Theory of Justice,* p. 334, italics added.

duty is necessary, but departs from the strong reading in rejecting that this is all that justice requires.

Although I believe the case can be made for the exclusive reading of the institutional duty, my discussion here need presume only the inclusive reading. That is, attending to institutions is a necessary requirement of acting justly even if this alone is not sufficient in terms of discharging one's duty of justice. Option (d) holds that whatever else climate justice can require of us personally (e.g., to reduce our direct emissions), it requires that we do our part to bring about better social arrangements to regulate indirect emissions. Whatever else we may achieve through personal action, we fail our duty of justice when we do not do our share to establish just arrangements when they are absent.

As we saw above, personal action or non-institutional duty is inadequate to the cause of climatic justice. It is either ineffectual (and overly demanding) or insufficiently demanding. Since total emission is in large part indirect and collective, effectively reducing emissions will require institutional rather than simply personal responses. That is, we will need better environmental laws and policies, stricter regulation of industry and agriculture, new infrastructure in areas like public transportation and urban design, governmental support for developing and installing clean energy alternatives and the like.

Consider, again, the collective and indirect nature of CO_2 emissions. Per capita emission is 18.4 metric tonnes for Australia, and 17.3 tonnes for the US, for example. On the other hand, China emits 5.6 metric tonnes per capita and India 1.6 tonnes.[24] Now it is hardly a stretch to say that the individual American or Australian on average directly emits more CO_2 personally than the average Chinese or Indian, due to their direct personal consumption of resources, energy, life-style (driving, traveling for work, household heating and cooling, operation of household appliances, vacation related travels etc.) and so on. But as noted, it is also obvious that personal direct consumption for the average American does not amount to 17.3 tonnes. The per capita emissions takes account of the total amount emitted due to joint human activities, and much of this is not reducible to the sum of direct personal consumption. For example, heating and cooling of public spaces and other shared spaces (office buildings), industrial and agricultural activities, emissions standards for cars and appliances and the country's means of producing energy are contributors to CO_2 production that are largely outside the control of individuals personally. Yet these factors are hugely responsible for the high emissions per capita in developed countries.

Thus an individual's reduction of their own personal carbon footprint – by driving less, recycling more, reducing personal energy consumption, reusing

---

[24] Figures from World Bank, http://data.worldbank.org/indicator/EN.ATM.CO2E.PC, last accessed 27 October 2014.

and so on – will not have significant impact on their country's per capita emission if its methods of energy production and distribution and collective usage remain unchanged, if its industrial and agricultural practices and policies stay constant, if there aren't alternatives in place like better public transportation, urban planning and the like. Thus significant emissions reduction cannot be achieved through personal action alone – such as recycling and reducing consumption – but must require collective action. Real change can be brought on only through infrastructural changes, revisions and institutional reforms. These infrastructural changes will require thus a focus on institutions rather than on personal conduct.

Institutional changes then are necessary to address indirect emissions. We might treat this as a kind of public goods problem: the common good of emissions reduction cannot be obtained through personal choices alone, not even through the combined effects of these choices, but can be secured only through coordinated public action. Besides efficacy, institutional responses, as already mentioned, establish personal expectations and clarify individuals' personal responsibilities. A society's burden of climatic justice has to be assigned and allocated among its members. There is no pre-institutional or *a priori* just way of doing this. In any society, there will be a range of possible institutional sets that can satisfy the requirements of just allocation of climate change burdens. Having a just regulatory regime in place will establish publicly what citizens' duties of climatic justice are. It solves the indeterminacy problem that option (c) in one of its forms faces.

Very importantly, the institutional focus signals that the pursuit of justice is a joint commitment and a collective enterprise. It is a responsibility of society to bring about justice, and the duties of justice individuals have in this regard are duties they have as members of a collective. Importantly, the institutional approach also understands justice to be a structural problem rather than a problem wholly reducible to personal conduct as such, and so is a problem that, on account of its origin, needs institutional rather than merely personal responses. Understood as a collective project, each person has to publicly acknowledge their obligations of justice and to comply with them. Yet, as I suggested earlier, in the absence of actual arrangements, it is not easy, if impossible, to know what each of our specific duties are. As Rawls writes, 'a person's obligations and duties presuppose a moral conception of institutions and therefore that the content of just institutions must be defined before the requirements for individuals can be set out'.[25] Thus there is a natural duty of justice for Rawls that individuals establish just arrangements when these are lacking.

[25] Rawls, *A Theory of Justice*, p. 110.

We may each understand that we have a moral obligation to reduce, recycle and reuse, taking these to be duties we have independently of background institutions. But this is an imperfect duty absent further institutional specification. And moreover as noted, personal efforts to act on the three Rs in the absence of institutional reform will either be insufficient (if what one does what one reasonably can) or overly demanding and still futile (if one is to go all out to achieve some ideal of just per capita emission personally).

It is well known that an institutional focus helps mediate personal duties and coordinate individual effort. Also, without institutional changes, some adjustments to personal habits that just emissions might require will be unreasonably costly. Institutional changes can mitigate the cost of personal actions needed to slow climate change. For instance, the moral demand that we drive less may be for some individuals especially personally costly without background structural and societal changes, such as the provision of alternative modes of travel.[26] My focus here is less on this mediating effect and coordinating roles of institution, although this is also an important case for the institutional approach. I am concerned here less with the benefits of focusing on institutions and more on its necessity.

My basic point is that climate change justice requires governmental action. This might seem a rather obvious remark. But it is worth reiterating since so much of our environmental movement and education focuses on personal conduct. School children are taught to reduce, recycle and reuse. Public campaigns likewise tend to focus on personal conduct and choices. This is all valuable, and public awareness and education on how each can contribute by reducing and changing lifestyle choices must continue. But what should be more explicitly on the environmentalist educational agenda is the encouragement of more individual political engagement and action. Real change can come only with better institutions. The task of justice has to be ultimately a political one – that of urging governments to take the demands of justice seriously.

But this chapter hopes to also make a less obvious point, which is to support the institutional ideal that the most important duty of justice where just arrangements are absent is to help bring them about. Some critics have said that this confuses a means for an end. But the problem of climate change justice very well illustrates the inadequacy of personal efforts to promote justice in the absence of institutional changes. Far from being 'fundamentalist', the view that we must focus on institutions acknowledges that justice is a collective and political problem and quest. If the source of injustice is collective and structural in nature (due for

---

[26] See here Meyer and Sanklecha, 'Individual expectation and climate justice'.

example to high collective impersonal emissions), then the response must be similarly collective and structural – the response to injustice must necessarily be institutional in its focus.

What should individuals do that will help bring about just climatic institutions? The short answer is more political action and not just personal acts of reducing, recycling and reusing. Voting for the right government based on their climatic agenda, lobbying for better regulation and restrictions on industrial and agricultural activities, calling out legislators who do not support laws that will help reduce collective emissions, putting pressure on governments to adopt global agreements on climate change, inciting educational reforms to promote understanding of climate change science, and so on.

It is correct that the formal claim I am defending here – that persons should aim to bring about better arrangements – does not itself specify what individuals should each do. What it does, however, is to identify the target and goal. It is a somewhat open question still, admittedly, what each should do with regard to this goal of bringing about just arrangements. But the structural point says something significant: it says that each of our obligations of climate justice is not discharged simply by our making personal choices within the terms of our current institutional arrangements. What is needed is institutional reform, and we are to do our part, in concert with others, to bring this about. Responding to climate change requires more than personal actions within existing laws; it will require that we consider laws as they might be.

Conclusion

My claim is not that justice does not require personal action or non-institutional duty at all. It might very well be that the personal actions are better understood as acts of beneficence rather than acts of justice on the institutions approach. But my point here, more modestly, is only that personal actions, whether they are required by justice or not, are not equal to the cause of justice. The most important duty of justice is still that of creating just arrangements when they don't exist. Justice requires at the very least efforts by individuals to create just arrangements, efforts that are not morally substitutable by personal actions of other kinds.

In the case of climate justice, more so than reducing their personal emissions through individual lifestyle changes, citizens of developed countries have the important responsibility of justice to press their governments to establish a just regulatory framework. They have the responsibility of justice to take the necessary political action to bring about the appropriate institutional changes. As Walter Sinnott-Armstrong writes:

We should not think that we do enough simply by buying fuel-efficient cars, insulating our houses, and setting up a windmill to make our own electricity. That is all wonderful, but it does little or nothing to stop global warming and also does not fulfill our real obligations, which are to get governments to do their job to prevent the disaster of excessive global warming.[27]

Climate change justice requires political engagement and not, or certainly not just, personal action.

---

[27] Sinnott-Armstrong, "'It's not *my* fault", p. 344.

8 Acts, omissions, emissions

Garrett Cullity

On at least four grounds, it can be argued that we ought to be reducing our carbon emissions in order to mitigate the impact of harmful climate change. The first is an argument from collective prudence: we ought to avert harm that we will otherwise suffer. The other three are moral arguments. There is an argument from intragenerational harm: the emissions produced by some of those now living will harm other members of the current global population. There is an argument from intergenerational fairness: our generation is greedily using up a greater share of the Earth's resources, including its atmospheric carbon-carrying resources, than we are willing to leave to those who will come after us. And there is also an argument from intergenerational harm, which simply insists that we ought not to act now in a way that will harm those who will live later.[1]

When these arguments are made, who are 'we'? There are various candidates – the whole global population across time, all those now alive, the high-emitting subset of the world's current population, all the high emitters who will ever live – and the arguments just listed seem to target different ones. However, you and I are members of all of those groups. If those arguments are forceful (as I think they are), then what moral implications does that carry for our actions as individuals?

Difference-making and participatory derivations

There are apparently cases in which, although a group is not doing what it ought to do, no individual member of the group acts wrongly. That is true even when the 'ought' is moral. For example, when one group has a special duty to another – say, a duty to apologise for past wrongs – it may not be possible for an individual member to do what the group ought to do (what may be needed is an apology by the Australian government, not by an individual Australian citizen), and it may be supererogatory for any individual to get the group to do what it ought.[2]

---

[1] On the 'non-identity problem' for this claim, see footnote 22, this chapter.
[2] For further discussion, see H. Lawford-Smith, 'The feasibility of collectives' actions', *Australasian Journal of Philosophy*, 90 (2012), 53–67.

According to some writers, anthropogenic climate change is another example of group wrongdoing without individual wrongdoing. They argue that here, the actions of one individual make no difference to global climate change. So even if all four arguments establish that *we* ought collectively to reduce our carbon emissions, I do not act morally wrongly in not reducing mine.[3]

However, that is too fast. It is probably true that my carbon-emitting actions make no difference to anyone else's welfare. But it does not follow that they are not morally wrong. That inference is blocked in two ways. First, an action can be wrong not because of the actual harm it does but because of the expected harm associated with it – that is, the sum arrived at by multiplying the value of each possible outcome of the action by its probability. I shall return to that issue. But second, even when your action does not even have an expectation of harm, it can still be wrong because of its relationship to what we do together. There can be what I shall call a 'participatory derivation' of the wrongness of your action from facts about what we ought to do and your actual or potential participation in our doing it, independently of the effects of your individual action or inaction. This is one of the things that may be meant when it is claimed that I bear a 'mediated' responsibility for climate change – a responsibility mediated by my membership of a group that either causes it or can prevent it.[4] The main aim in what follows will be to see whether this thought stands up to closer examination.

Participatory derivations can be distinguished into two broad kinds, positive and negative. In cases of the positive kind, a group ought to be doing something, it is doing it, and I ought to join in, bearing my share of the burden of the collective action. Under exactly what conditions that is true is disputed; but most moral philosophers recognise cases of 'free riding' on others' production of public goods that are morally wrong.[5] If we ought to cooperate in digging a well for our village, and I leave it to others to dig it but then help myself to the water afterwards, I am behaving wrongly. To justify not joining in, I would need to argue that we ought not to dig the well; once I accept that we ought to do

---

[3] W. Sinnott-Armstrong, '"It's not *my* fault": global warming and individual moral obligations', in W. Sinnott-Armstrong and R. B. Howarth (eds.), *Perspectives on Climate Change: Science, Economics, Politics, Ethics*. Advances in the Economics of Environmental Resources (Amsterdam: Elsevier, 2005) vol. V, pp. 285–307; B. Johnson, 'Ethical obligations in a tragedy of the commons', *Environmental Values*, 12 (2003), 271–87. Cf. D. Killoren and B. Williams, 'Group agency and overdetermination', *Ethical Theory and Moral Practice*, 16 (2013), 295–307; J. Sandberg, 'My emissions make no difference: climate change and the argument from inconsequentialism', *Environmental Ethics*, 33 (2011), 229–48.

[4] R. Attfield, 'Mediated responsibilities, global warming, and the scope of ethics', *Journal of Social Philosophy*, 40 (2009), 225–36.

[5] For further discussion, and a survey of relevant literature, see G. Cullity, 'Moral free riding', *Philosophy and Public Affairs*, 24/1 (1995), 3–34, and 'Public goods and fairness', *Australasian Journal of Philosophy*, 86 (2008), 1–21.

so, taking a free ride is wrong.⁶ This is wrong because it is unfair. The free-rider arrogates a special privilege to himself – the privilege of taking a good without paying – while relying for the existence of that good on others' willingness to pay. And this can be true when the free-rider harms no-one. If I sneak into a theatre without paying, I might impede no other theatregoer's enjoyment (indeed, I might enhance it, if I clap at the right times). The moral complaint against me is not that my action harms anyone. It is that I am failing to contribute what is required from similarly situated people if the group is to function as it ought.

Participatory derivations can also take a negative form. In these cases, there is something we ought *not* to be doing, and I am joining in our doing it. Suppose a gang has dragged someone into a building to assault him, and I am asked to guard the door and shout a warning if the police arrive. If I agree, I become complicit in the assault: I make myself a member (even if a minor one) of the gang that perpetrates it. But here, too, my actions may make no actual or even expected difference to the victim's welfare. Perhaps no police arrive; I shout no warning; had I declined, someone else would have taken my place. Nonetheless, I remain morally accountable for joining the group that perpetrates the assault. The victim can rightly complain that I was one of the gang that assaulted him.⁷

In the rest of this chapter, I want to examine whether a participatory derivation of either of these two types can support the conclusion that individuals' energy-consuming activities are wrong. Both need to be considered. Not constraining my own personal carbon emissions might be characterised either as a failure to join in what we collectively ought to be doing (moderating our emissions) or joining in what we ought not to be doing (emitting at levels that cause climate change). I shall identify what I think is the strongest such derivation; but my conclusion will be that it is hard to see how to make it completely convincing.

Wrongness

I should first spell out how *wrongness* is being thought of in this discussion. There are different ways of thinking of the force of the complaint that an action is morally wrong and why it matters. Although it can be tempting to ask which

---

⁶ Arguing that we ought not to be digging the well would be the start of a case for the conclusion that I am not acting wrongly, but not the end of it. If the community has fair and inclusive procedures for collective decision-making, then I could be morally required to respect the decisions it has reached, despite disagreeing with them.

⁷ I am not claiming that if the gang acts wrongly and I am part of the gang, then that is sufficient to make my action of joining the gang morally wrong. If I guard the door under duress, I am part of the gang, but the duress involved in making me join may have an excusing force.

one is correct, what seems more constructive is to identify the differences between them, and specify the one you are following.

A full list of conceptions of wrongness would have many members, no doubt, but three prominent ones are these. On an optimisation conception, wrong action is sub-optimal action: the force of complaining that an action is wrong is that it is not as good as some alternative. On a compliance conception, morality consists in a set of proscriptions, and for an action to be wrong is for it to be proscribed by that set: the force of complaining that an action is wrong is that it fails to comply with the commands imposed by morality. On what we can call 'adequate reasons' conceptions, wrong action is action that fails to respond adequately to the other-regarding reasons bearing on it. On a view of this third kind, the force of complaining that an action is wrong is that it treats others unjustifiably.

Adequate reasons conceptions of wrongness are a family of different views, which offer different ways of specifying what it is to respond 'adequately' and what makes a reason 'other-regarding'. The 'others' in question might be restricted to rational agents, human beings or sentient creatures, or extended to include the non-sentient environment or precious objects in general. And the 'adequacy' of my response might be specified by reference simply to the relative strengths of the reasons for and against the action;[8] to whether the action calls for reactive attitudes such as resentment, indignation and blame;[9] to a publicity condition, concerning the reasons that must be recognised as prevailing if our interaction is to be governed by the exchange of reasons rather than coercion;[10] or to whether the action meets the demands of second-person accountability we are entitled to address to each other.[11]

Here, I shall prescind from such differences, important though they are. They are rival ways of filling out a core idea – that the relations in which we stand to others provide us with reasons and that an important concern, expressible by talking about wrongness, is that our treatment of others should be justifiable in the light of those reasons. My question about wrongness will be whether our individual actions meet that general demand. (I shall use 'moral requirements' to refer to what it is morally wrong not to do.) I began by pointing out some of the ways in which a harmless action can fail to meet that demand. We should now examine them in more detail.

---

[8] This is a standard reading of W. D. Ross's view of wrongness: see for example P. Stratton-Lake, 'Introduction', in W. D. Ross, *The Right and the Good* (Oxford: Clarendon Press, 2002), ix–lviii.
[9] See e.g. G. Watson, *Agency and Answerability* (Oxford: Clarendon Press, 2004), esp. ch. 8.
[10] T. M. Scanlon's view, in *What We Owe to Each Other* (Cambridge, MA: Harvard University Press, 1998), ch. 5, belongs to this type.
[11] See e.g. S. Darwall, *The Second-Person Standpoint: Morality, Respect, and Accountability* (Cambridge, MA: Harvard University Press, 2006), esp. ch. 5.

Expected harm and benefit

The expectation of harm associated with an actually harmless action can make it wrong: reckless and negligent actions illustrate that. This provokes a first line of reply to the argument that an individual's carbon-emitting action makes no difference and therefore cannot be wrong.[12] According to the best understanding of the current science, the carbon we put into the atmosphere already causes harm and will continue to do so.[13] Increasing global temperatures are correlated with an increasing incidence of extreme weather events – bushfires, floods, storms, droughts and heat waves – which cause death, injury and distress. It is true that the causal structure of weather systems is too chaotic to allow us to attribute particular events of those kinds to particular emissions, and that some emissions may well prevent such events. But there tend to be more of those events as the Earth gets warmer, and putting more carbon into the atmosphere tends to make it warmer. Given this correlation, my emissions, by adding carbon molecules to the atmosphere, have some probability of triggering a harmful event of this kind, and it is greater than the probability of preventing one. This probability is very small, but the harm if such an event is triggered is very large. So there is some number of deaths that represents the expected harm produced by my emissions. One attempt to quantify this puts it at one to two deaths over my lifetime; another, at shortening others' lives by two to three days per year of emissions.[14] It cannot be inferred from this that I cause that amount of harm.[15] I probably cause no harm. But that does not exonerate me, if the expected harm is significant.

However, deriving a conclusion about wrongness from this line of thought is not straightforward. Consider an analogy. Unsurprisingly, there is a correlation between ambient noise levels and the incidence of aggressive behaviour.[16] There is therefore some small probability that my contribution to the overall level of ambient noise will trigger a violent incident. I could reduce that probability by being quieter. Indeed, I could reduce it to zero, 'offsetting' my contribution by quietening some other significant source of noise. However, it

---

[12] For clear expositions of this line of reply, see C. Morgan-Knapp and C. Goodman, 'Consequentialism, climate harm and individual obligations', *Ethical Theory and Moral Practice* (2014); and J. Broome, 'The public and private morality of climate change', *The Tanner Lectures on Human Values*, University of Michigan, 16 March 2012.

[13] See the Intergovernmental Panel on Climate Change (IPCC), *Climate Change 2014: Impacts, Adaptation and Vulnerability*, www.ipcc.ch/report/ar5/wg2, last accessed 10 October 2014.

[14] Broome, 'The public and private morality of climate change'; J. Nolt, 'How harmful are the average American's greenhouse gas emissions?', *Ethics, Policy and Environment*, 14/1 (2011), 3–10, p. 9.

[15] As Broome, 'The Public and Private Morality of Climate Change' and Nolt, 'How harmful are the average American's greenhouse gas emissions?' both do.

[16] This correlation is found in those with aggressive predispositions. See S. Cohen and S. Spacapan, 'The social psychology of noise', in D. M. Jones and A. J. Chapman (eds.), *Noise and Society* (Chichester: John Wiley, 1984), 221–45, see pp. 227–31.

is hard to take seriously the idea that I act wrongly in not doing so. Although we have no figures, the expectation of harm seems too small in comparison to the effort it would take to reduce it, and the causal contribution to harm too remote, to make it a reasonable object of complaint.

That does not refute the case for thinking that the associated expectation of harm makes individuals' carbon-emission behaviour wrong. It is easy enough to see that that case is stronger than the corresponding one for ambient noise-making, in all three of the dimensions that need to be compared. The expectation of harm is greater. The associated cost is modest: paying $550 will offset 25 tonnes of personal carbon dioxide emissions, we are told.[17] And the causal contribution is less remote. When violence is triggered by ambient noise, the causal contribution of the person who strikes the blows dwarfs that of a background noise-maker. But when carbon emissions cause a hurricane, there is no intermediate harmful agent.

However, while it is obvious that the case is stronger, is it strong enough? Once we have quantified the expected harm and the cost of offsetting it, there are three more things to do. We need to compare the relative sizes of those two magnitudes. We need to assess the centrality or remoteness of the causal contribution of current carbon emissions to future harm. For example, if an elderly person dies in a future heat wave, we face the question how to apportion the relative contributions of past carbon emitters, the lack of precautions by the person and his relatives who know about the hotter environment, and the failure of health services to adapt. And having made the first comparison and the second assessment, we need to relate them to each other. It is hard to know how to do those things. One possibility would be to look for a similarly structured case in which factors of the same magnitude were associated with obviously wrong action; but I cannot see how to do that.

Another issue adds to the complexity. Suppose you do offset all your emissions, and eradicate the expectation of harm associated with your actions.[18] You now face a further question. You have important other-regarding reasons not just to avoid harming other people, but to benefit them. Failing to act in a way that has a chance of saving other people's lives at comparatively small cost to you can be wrong too – even when the threat has been caused by someone else. So if you are sufficiently convinced by the argument from expected harm to offset your own emissions, you need next to ask yourself why you are not going further and offsetting other people's. Arguably, the reason you have to perform actions with an expectation of preventing harm is weaker than the reason not to perform actions with the corresponding expectation of harm; the cost of offsetting other

---

[17] Source: www.climatefriendly.com.
[18] You do not thereby eradicate the expectation of harm associated with your emitting actions; but the expectation of harm associated with your emitting and offsetting actions taken together is zero.

people's emissions as well is greater than the cost of only offsetting your own; and if you do not, your causal contribution to future harms that result from other people's emissions is more remote than to those resulting from your own. If so, the case for the wrongness of not doing this is weaker. But until we know how to measure the bearing that these three factors have relative to each other on the wrongness of action, we cannot conclude that the latter case is too weak to establish a conclusion of wrongness – nor that the former case is strong enough.

Not joining in what we should be doing

An argument that individual emissions are wrong because of the associated expectation of harm is a non-participatory derivation. It applies to me independently of my membership of any group. If I inhabited a world in which the same atmospheric composition had not been caused by humans, and I somehow faced carbon emission choices all on my own, the same considerations would apply. But now let us examine the prospects for a participatory derivation. We can start with the first, positive kind.

In this kind of case, my action is wrong because it is a failure to join in an action we ought to be performing. Free-riding was cited to illustrate this: this is wrong because it treats unfairly those who do join in – even if it does not reduce anyone's welfare.

Notice that this involves a derivation from what we ought prudentially to do to what I am required morally to do. When we ought to dig a well, the 'ought' need not be moral: it suffices that the action furthers our collective interests. But the requirement on me is a moral requirement, of fairness.

Positive participatory derivations extend beyond cases of free-riding, however.[19] Suppose someone is drowning, and eight of us standing on the beach could save him by rowing a lifeboat together. Then we ought to do that. This time, the 'ought' is moral; but again, there is a participatory derivation. If I decline to join in, leaving the other seven to do the work, then I am acting wrongly. Here too, I cannot exonerate myself by pointing out that the victim will still be saved and that, although the others will have to work a bit harder without my participation, this will not harm them in any significant way. The complaint against me is not that my behaviour makes a significant difference to anyone's welfare. It is, again, that I am failing to contribute to what we together ought to be doing. It is only because the others do not think and act like me that the victim is saved. However, I am not free-riding – helping myself to a good

---

[19] Nor does it seem that all cases in which free-riding is unfair follow this derivational form. If I help myself to the goods produced by a gang's protection racket without contributing to the efforts of the gang, that might treat the other gangsters unfairly. But it would not be because I ought to be joining in those efforts. For that to be true, it would need to be the case that the gang ought to be imposing its protection racket, which is false.

produced by the group. And here, it seems unsatisfactory to explain the wrongness of my action solely by saying that I treat the rescuers unfairly. An account of who I am mistreating cannot just mention the rescuers: it must include the victim. The moral complaint against me is best put disjunctively. I am either treating the victim callously, failing to accept that his life must be saved; or treating the other rescuers unfairly, leaving them to do the work while arrogating to myself the privilege of not contributing.

Apparently, then, positive participatory derivations cover a range of different cases. More work would be needed to settle exactly how extensive this range is. Does it include cases in which a group action confers a benefit on me without my having actively sought it? And just what determines when a group that ought to be doing something includes me? However, we do not need to address those questions here. Before we reach them, there is already a powerful obstacle to generating moral requirements on individuals in relation to climate change in this way.

This is simply that a derivation of this form does not generate unilateral moral requirements. In a positive participatory derivation, I am required *to join in* what we collectively ought to be doing. If we are not doing it, then there is no collective action for me to join in, and therefore no basis for a complaint that I am unfairly failing to do so. If we ought to be digging the well but are not, I am not morally required to make a hole in the ground that corresponds to my share of the work of actually digging a well together; if we ought to be rescuing the drowning person but are not, I am not morally required to get in the lifeboat by myself and row in a circle.

The problem for a positive participatory derivation in the case of climate change, then, is simple. The global population ought to be regulating our economy to contain carbon emissions. But we are not. So there is no collective action which I am failing to join. Unilaterally constraining my own carbon emissions is like sitting in the lifeboat by myself and rowing in a circle.

There is a reply, however. When a large-scale problem requires a large-scale collective action that is not being performed, smaller-scale collective actions may still address part of the overall problem. If so, a positive participatory derivation can still generate a moral requirement to join in the smaller-scale action. Perhaps many drowning people need to be saved by launching many lifeboats, but there are only enough willing rowers to launch one. Then I cannot join a grand collective action that saves everyone; but I should still join the crew of the one boat that will save some.

The application to carbon emissions is this. I cannot join effective worldwide action on the scale needed to arrest climate change. But I can join the many individuals worldwide who are living carbon-neutral lives. Their collective action will not avert the dangers of climate change fully. But it makes the impact less than it would otherwise be, so I should join in, rather than leaving that work to them.

In fact, there are four distinguishable arguments of this type to consider. (1) Those who are leading carbon-neutral lives produce a public good. They ought to do that for reasons of collective prudence, and since they confer a benefit on me, I am unfairly free-riding if I decline to join them. (2) They are preventing harm to vulnerable people, both now and in the future. They ought to do that for reasons of beneficence, and my refusal to join in is wrong in the same way as refusing to join the lifeboat crew would be. (3) While the high-emitting members of the world's population use a greater share of the Earth's resources than they are leaving to future generations, this group does not. This group's actions do not leave the world in worse shape than they found it: they ought to perform those actions for reasons of fairness. There is a derivative requirement of fairness on me to join in, rather than leaving to others the work of doing what we ought. (4) The large-scale action that is needed to address climate change is effective regulation of the global economy. When that is not happening, we face the question of what we ought to do to encourage it. If the lifeboat is not being launched, I should not simply walk off the beach; I should see whether I can gather a group to launch it. And if rowing the lifeboat in a circle will shame the others into joining in, then I *should* do that. By leading carbon-neutral lives, we can send a political signal that makes effective global regulation likelier. So we ought to do that, and I ought to join in rather than leaving the work to others.

However, each of these arguments faces problems. According to (1), in benefiting from others' self-constraint I am free-riding. But as most discussions of free-riding recognise, refusing to help produce a public good is only unfair when the benefit you receive exceeds the cost you are being asked to pay.[20] A scheme of collective prudence from which no contributor gains a net benefit ought not to be supported at all; if some do while others do not, the contributory demand on the latter is unfair. And while working out the benefit that a rich-world individual receives from others' self-constraint would be complicated, it is hard to believe that high-emitting individuals, cushioned as they are from the worst effects of climate change, are actually receiving a benefit through others' voluntary self-constraint which is greater than the cost of imposing that constraint on themselves.[21]

---

[20] See e.g., R. Arneson, 'The principle of fairness and free-rider problems', *Ethics*, 92/4 (1982), 616–33, p. 623. Complexities arise here concerning the bundling of public goods. It can apparently be reasonable to require me to pay for some bundles of goods even though not every good in the bundle is beneficial to me.

[21] This would require a three-step calculation. The first step would be to assess how much the carbon-neutral group would have emitted had they not imposed this constraint on themselves. For this, one would need data about people's incomes, their emissions and whether they report that they are constraining those emissions out of concern for climate change or not. One could then estimate the amount by which (reported) self-constraint reduces emissions, on average, in people on a given income. Given that estimate, the second step would be to determine the difference their self-constraint makes to global temperatures. The third would be to correlate that temperature difference with premature deaths and other effects on people's welfare.

Consider argument (3) next. This faces a different problem. It is true that the group of people leading carbon-neutral lives can correctly say, 'We do not leave the world in worse shape than we found it'. It is others who mistreat those who will inherit a degraded world: this group does not. But this does not ground a positive participatory derivation. There may be reasons for me not to produce emissions that carry an expectation of harm (as discussed above); and there may be reasons for me not to join in collective actions of harming (to be discussed below); but the complaint to make against me is not that I am unfairly leaving the burden of not mistreating future generations to others. Compare: in a society pervaded by racism, the complaint against racists is that they are mistreating people; it is not that they are unfairly leaving the burden of not mistreating people to non-racists.

Argument (4) makes an important point. The absence of effective global action prompts the question what we are willing to do to secure it. However, it is fanciful to think that simply reducing energy consumption itself sends an effective political signal. Joining a political lobbying movement can be a way of participating in sending such signals; simply turning the lights off is not. Governments do not treat patterns of energy usage as an indicator of people's opinions concerning climate action. Having said this, persuasive advocacy does often involve modelling the behaviour you are trying to encourage others to adopt. That provides a rationale for some advocacy groups to require carbon-neutrality of their members. So if it is *those* advocacy groups you ought to join, then a positive participatory derivation will generate a requirement of fairness on you to bear the same burden as other members. However, a compelling argument of this form would require that the most effective kind of advocacy group, and therefore the one I ought to join, makes this requirement of its members. And that seems difficult to establish.

The most serious of these arguments is (2). This maintains that the collective action of self-constrainers makes others (vulnerable people who are not members of that group) significantly better off than they would otherwise be. It asserts a moral requirement of beneficence – a requirement of the kind that is generated when someone can be greatly benefited at small cost to the beneficiary. It is true that 'greatly benefited' and 'small cost' are both scalar and vague. So requirements of beneficence come in degrees, and admit of indeterminacies. But there are clear cases: preventing someone's death by giving up money that you will not miss is morally required. This remains so when the benefactor is a group. If, by acting together, we can prevent someone's death by each giving up money we will not miss, then that can be morally required. But moderating global warming will prevent the deaths of vulnerable people, and the cost of offsetting to achieve carbon-neutrality is money a rich-world individual will not miss. So the collective action is morally required, and if so, the earlier disjunctive complaint can be made against someone who fails to join in. Either you think preventing people's deaths does not matter, or you are unfairly leaving the work of doing that to others.

To evaluate this argument, we need to appreciate its relationship to the non-participatory argument discussed earlier ('Expected harm and benefit'). That concerned the expected value of my individual action. This one concerns the expected value of our collective action. This is greater, because there are more of us. The argument then seeks to derive a requirement on me to join in which is independent of the expected value of my doing so: it is a requirement of fairness. To appreciate the structure of the argument, consider its application to another example. A large group is co-operating to free a trapped person by passing rubble down a very long human chain. If I refuse to join in, the complaint against me is not that it makes a significant difference to anyone: it is that it is unfair. The difference *we* make is why we ought to act; we are only doing so through the willingness of each individual to contribute despite making no difference; in being unwilling to cooperate on the same terms, I am being unfair.

Notice also that when we cooperate in this way, the costs each participant can reasonably be required to bear do not seem to be diluted by the size of the group. If someone is drowning and I am the only swimmer, then plausibly I should swim to the limit of my safe capacity to help him – I am not required to go beyond that. But if twenty of us could cooperate by swimming in relays, *each* of us should be willing to swim to the limit of his safe capacity. It is not as though, given the maximum I could be required to do on my own, I can divide that by twenty to generate the maximum I can be required to do when there are twenty of us.

This makes an argument of this form more powerful than the earlier non-participatory one. Now the expected value we need to calculate is that of our collective action, whose effects are much more significant than those of my individual action. But the cost I can be required to bear in cooperating with others seems as great as it would be if I were acting alone. However, sustaining this argument still requires making the same three-way comparison we struggled with earlier, only this time applied to the group – a comparison between the expected value of the action, the cost (to each member) of performing it, and the centrality or remoteness of its causal contribution. And although the case for thinking that the group stands under a moral requirement is stronger than it was for an individual, I am still unable to demonstrate that it succeeds. This is the strongest argument we have considered so far, but I do not know how to show that it is successful.

Joining in what we should not be doing

Although the argument just discussed needs to be taken seriously, it invites an accusation of moral complacency. Our relationship to future generations is not one of potential benefactors who should band together to bestow a *benefit* on them: rather, we are actively *harming* them, and I am participating in the collective harm.[22] This thought is emphasised in a participatory derivation of the second, negative, type.

I introduced this with the example of guarding the door for the gang. When I do that, I might make no difference to how much harm the gang inflicts on its victim. Why can I not infer that I have done nothing morally wrong? The answer is that in agreeing to guard the door, I participate in the collective action of the gang. The gang's action is wrong: it violates both the welfare and the human dignity of its victim. It is by joining in that action that my own conduct is wrong. Just as the victim can ask the gangster who struck the blows, 'Why did you assault me?' he can ask me, 'Why did you join the assault on me?' My having no adequate reply is what it is for my action to be wrong.

It is important to the force of this explanation that my participation amounts to more than just being a mereological part of what the gang does. Had the gangsters rigged my phone to a distant bomb, then in using the phone my action would be part of their attack. But I would not be directing myself towards the end of harming the victim, in the way that I do in agreeing to guard the door. When the gang members cooperate with each other to perform a collective action – an action of which the gang collectively is the agent – what makes that true is the interlocking structure of intentions and reasoning through which the members combine their agency.[23] The practical thought and activity of each member is responsive to that of the others, disposing them to make mutual adjustments of their individual contributions towards the group's achievement of its ends. In agreeing to guard the door, I contribute my own intentions to that structure. I am now disposed to shout a warning if I see the police, loudly enough to ensure that the others hear, in enough time to let them escape, to react if I am told to guard a different door, and so on. These dispositions may remain unactivated, but even if they do, they constitute a way in which my agency is structured towards supporting the end of the gang – the same end to which the other members' agency is structured, in a mutually responsive way. That end is the harm to the victim. This is why, when the victim's question is put to me – 'Why did you join in the assault?' – it is insufficient to reply that my joining in did not itself make a difference. My agency was structured towards the gang's

---

[22] There is a *de dicto* reading of that claim on which it is unaffected by the 'non-identity problem' – the point that our actions make a difference to which future individuals exist. Our actions do not cause the same individuals to be worse off; they cause one set of future individuals to be worse off than another set would have been. To most of us, *that* seems to be enough to be able to make current actions wrong because of their future effects. For discussion, see C. Hare, 'Voices from another world: must we respect the interests of people who do not, and will never, exist?', *Ethics*, 117 (2007), 498–523.

[23] See M. Bratman, 'Shared cooperative activity', *Philosophical Review* 101 (1992), 327–40; M. Gilbert, *Living Together: Rationality, Sociality, and Obligation* (Lanham, MD: Rowman and Littlefield, 1996).

end, of harming him.[24] That is an end there are powerful other-regarding reasons not to adopt. Lacking an adequate justification for that is what it is for my joining in to be wrong.

But if all negative participatory derivations must follow this form, then its application to climate change is problematic. The gang is an instance of *collusion*; anthropogenic climate change is not. There is no grand global agent coordinating its actions towards the deliberate end of worsening the climate. My own dispositions of thought and action are not structured, in combination with others', towards the production of that outcome. It is true that my own energy-consuming activities are a mereological part of our collective greenhouse gas emissions. But that does not mean I am participating in wrongdoing in the fuller sense just described – directing my own agency, through the coordination of my dispositions to act and reason about action, towards an immoral end which is the focus for interlocking contributions to the sharing of agency.

So my contributions to climate change are not wrong in that collusive way. But now we must broaden our view. Not all negative participatory derivations – those in which my action is wrong because it is a way of joining in what we should not collectively be doing – have the same collusive structure. There are other clear cases – for example, cases of collective negligence or recklessness. Suppose a group of us runs a business that discharges toxic waste into the water supply, and people are poisoned as a result. Our action need not be intentionally coordinated towards the end of poisoning people in order to be wrong. An action – collective or individual – can be wrong not because of what it aims at but what it fails to aim at. Collective action can be negligent or reckless; and when it is, the derivation of the wrongness of acts of individual participation in the collective wrongdoing is just as plausible as before. The individual business partners are structuring their own actions towards being part of the collective action. The reasons that make the collective action wrong are reasons for the individuals not to structure their actions towards it.

This might then seem to give us the model for a negative participatory derivation that applies to climate-affecting energy consumption. It is wrong for us collectively to pursue a course that negligently produces harmful climate change. Given this, my participating in this collective negligence, by myself indulging in the polluting activities that constitute it, is also wrong.

---

[24] My agency can be structured towards supporting this end of our collective action, even if it is not an end of my own. In guarding the door, my end need only be to earn some money – it need not matter to me whether the victim gets harmed. But in agreeing to act as I do, my agency is still structured towards supporting the attainment of that end by the group. My structuring my agency in this way, interdependently with the other gang members, forms part of the overall coordination of the group towards its aim. This is what makes it true that I am participating in the gang's attack, and makes me accountable to its victim.

However, this application is flawed too. The complaint against negligent agents is that they direct their agency in a way that fails to take adequate account of its potentially harmful effects. But such a complaint requires that there is a negligent agent to whose actions it applies. There can be a failure to direct agency as one should only where there is agency. But that is what is absent in the case of global climate-impairing behaviour. There is no global agent directing its agency in an intentionally structured way – no candidate object of criticism for failing to direct that agency in a way that takes proper account of its potential effects. So I am not participating in a negligent collective action.

From this, we cannot conclude that whenever a group does not constitute a collective agent it is morally innocent. Return to the case of the lifeboat. If the eight bystanders fail to communicate with each other and consequently fail to launch the boat, we have not performed a negligent collective action. We have performed no collective action, because we have not banded together to form a collective agent: we are just a random collection of individuals. But it is not as though that is morally unproblematic.[25] On the contrary, we ought to have constituted a collective agent, and our failure to do that is a moral failure. Moreover, if I could have done something to promote our cooperation and failed to do that, I have acted wrongly.

Thus, no objection has been raised against seeing *that* as our moral predicament with respect to climate change. I face the question what I can do to promote cooperation on the scale required to address the problem. But that does not vindicate a participatory derivation. It derives a reason for me to act from the effects I can have on whether a group will act, not from the participatory relation I either bear or fail to bear to the actions of a group.

So: does that mean that negative participatory derivations require the existence of collective agents? No.[26] Suppose a mob is rioting in the centre of town, and I go there to join it. I do not myself break anything or harm anyone; I just run around and share the excitement while others do that. The mob might not constitute a collective agent: it might be just a crowd of individuals with no leadership and none of the coordination through which we can attribute intentions or deliberation to a group. That does not exonerate me. There is still a question I have to face, namely: 'Why did you join in?' Even though there was no coordinated group agency, there was still something the crowd did. I have aligned myself with its destructive actions, oriented myself towards its violent behaviour. That is something I can be challenged to account for – as I could, for that matter, if I had merely seen the rioters from my window and cheered them on.

---

[25] V. Held, 'Can a random collection of individuals be morally responsible?', *The Journal of Philosophy*, 67 (1970), 471–81.

[26] Nor do positive ones. The collection of theatre patrons do not have to constitute a collective agent in order for me to be unfairly free-riding when I sneak in without paying.

So here there is no collective agent, but the wrongness of my action still derives from the participatory relation I bear to what is done by the group. Why not take *this* as a model for the moral relationship individuals bear towards the large-scale phenomenon of climate change? The answer is that I fail to bear the orientational relationship to the larger group that is present in the example of the rioting crowd. If I were just inadvertently caught up in a riot happening around me, there would be no moral challenge to address. The challenge is: 'Why did you join in, aligning yourself in favour of the destruction?' The relation I bear to the uncoordinated actions through which climate change is occurring is not one of positive alignment; there is no joining in.

In examining negative participatory derivations, we began with cases of collusion towards a bad end. We then considered two further kinds of case, which meet weaker conditions. First, there are cases of collective negligence, which are not directed towards a bad end. Second, there are cases of joining in the bad action of a random collection of individuals, where there is no collective agent. I claimed that the moral objection in the first kind of case requires a misdirection of collective agency, and that in the second it requires a positive orientation towards a bad object. But neither of those things is present in the case of climate change.

That leaves us with the question whether there are further negative participatory derivations, meeting conditions that are weaker still – where there is neither a collective agent nor a positive orientation. Is it enough to make my action wrong simply that a collection of individuals together cause significant harm by performing an action of a certain type, I know this, and I knowingly perform an action of that type?[27] I think the onus is on someone who thinks that to explain why, and I cannot see what the explanation is. Sometimes, actions meeting that description are wrong; but when they are, the explanation comes from elsewhere. To give a single example: individual actions of littering are wrong, even though they involve neither participation in collective harmful agency nor a positive orientation towards a bad object. But we can explain their wrongness through a positive participatory derivation. An established social expectation of non-littering is sustained by many people; it produces the good of a less littered environment; so I ought to play my part in upholding it. Without that social expectation, that explanation would fail, and I cannot see how to supply a convincing alternative.

---

[27] Compare Parfit's principle (C12), D. Parfit, *Reasons and Persons* (Oxford: Clarendon Press, 1984), p. 81, which is invoked by Attfield, 'Mediated responsibilities, global warming, and the scope of ethics', pp. 228–9. But is Parfit's principle correct? If his 'Harmless torturers' (p. 80) are *colluding*, then each acts wrongly in the same way as someone who joins the gang. But what if they are not?

International applications

Our focus has been on looking for a participatory derivation of the wrongness of individual contributions to climate change. The discussion has an application to nations as well as to individual agents: I consider this briefly before closing.

Nation-states are agents. They do things, and what they do is determined by decision-making processes that involve deliberation about the reasons for and against the alternatives open to them. No doubt, the existence of national action, decision and deliberation reduces to facts about the individuals and institutions by which nations are metaphysically constituted. But that should not make us doubt that nations perform actions: the structural features that make you and me into reason-responsive agents are equally present at the level of nation-states.

So-called 'political realists' deny that national actions are candidates for moral assessment. The argument against that view is straightforward: since nations perform actions which are capable of responsiveness to reasons, it is intelligible to ask whether those actions take adequate account of the other-regarding reasons that bear upon them – whether they are morally right or wrong. Here, I simply register my view that the realists' counterarguments are unconvincing.[28] If so, the question arises: is there a participatory derivation of either of our two types that establishes a moral requirement on nations to take action on climate change?

With respect to negative participatory derivations, the answer again is No. Since there is no global collective agent that is either deliberately causing climate change or negligently producing it as the effect of some other collective action, nations are not participatory contributors to wrongful action of that kind. Here, their position is comparable to that of individuals.

However, with respect to a positive derivation, the position of nations is different. For while there may be no global collective action of the kind that is needed to regulate the world economy to contain carbon emissions, there is international action that is coordinated towards achieving such regulation. So individual nations ought at least to join in that larger collective action: leaving the work to others is unfair. Having said this, the most that can be derived from an argument of this kind is a requirement that one nation contributes its share of the effort that is actually being made internationally, not that it takes the lead by going further.

There is a requirement of leadership, but it has a non-participatory source. Participatory derivations are those in which moral requirements on agents derive from their actual or potential participation in a larger group,

---

[28] I discuss this issue further in G. Cullity, 'The moral, the personal, and the political', in I. Primoratz (ed.), *Politics and Morality* (Basingstoke: Palgrave Macmillan, 2007), pp. 54–75.

independently of the further effects of their own action or inaction. But the actions of nations do have significant further effects which bear on their moral assessment. A nation's actions are much likelier than those of an individual to affect what many others will do. Given this, there is a strong case for a moral requirement to exercise that influence positively.

Conclusion

I conclude that a participatory derivation of moral requirements on individuals in relation to climate change, either positive or negative, is difficult to sustain; and that when we consider its application to nations, it is limited. So the possibility noted at the outset remains open: perhaps our relation to climate change is indeed one of the cases where a group acts wrongly although no individual member does so.

In some ways that would be a disappointing conclusion to reach, since if it is true that may make it harder for us to motivate ourselves to address this problem properly. I have not quite reached it: I have not found a compelling argument that individual inaction is wrong; but there may be a better argument I have overlooked. However, I close with this thought. The question I have been discussing itself provokes a question. How important is it to establish whether an action is morally required – whether not performing it would be wrong? I have done nothing to cast doubt on the existence of good reasons to moderate one's own carbon emissions, as an expression of concern for what is happening to the world. After all, it would be bad enough to deserve the disdain of future generations for having worsened the world we leave them, whether or not we also deserve their blame for having wronged them.[29]

---

[29] This chapter has benefited greatly from the written comments of Aaron Maltais, Matthew Rendall, Anne Schwenkenbecher and Dan Weijers, and from audiences to whom earlier versions were presented at Adelaide, ANU, McGill, Melbourne and Wellington.

9 Individual responsibility for carbon emissions: is there anything wrong with overdetermining harm?

Christian Barry and Gerhard Øverland

Introduction

Climate change and other harmful large-scale processes challenge our understandings of individual responsibility.[1] People throughout the world suffer harms – severe shortfalls in health, civic status, or standard of living relative to the vital needs of human beings – as a result of physical processes to which many people appear to contribute. Climate change, polluted air and water, and the erosion of grasslands, for example, occur because a great many people emit carbon and pollutants, build excessively, enable their flocks to overgraze, or otherwise stress the environment. If a much smaller number of people engaged in these types of conduct, the harms in question would not occur, or would be substantially lessened. However, the conduct of any particular person (and, in the case of climate change, of even quite large numbers of people) could make no apparent difference to their occurrence. *My* carbon emissions (and quite possibly the carbon emissions of much larger groups of people dispersed throughout the world) may not make a difference to what happens to anyone. When the conduct of some agent does not make any apparent difference to the occurrence of harm, but this conduct is of a type that brings about harm because many people engage in it, we can call this agent an overdeterminer of that harm, and their conduct overdetermining conduct.[2]

What is the moral status of overdetermining harm? Four questions lurk within this broad one. First, are there moral reasons against becoming an overdeterminer

---

[1] Earlier versions of this chapter were presented at seminars at the Australian National University, University of Oslo and Centre for Applied Philosophy and Public Ethics. We are very grateful for comments received on those occasions from Geoff Brennan, Garrett Cullity, Keith Dowding, Bob Goodin, Frank Jackson, Seth Lazar, Jonathan Schaffer, Chad Lee-Stronach, R. J. Leland, Jeremy Moss, Ingmar Persson, Alex Sandgren, Anne Schwenkenbecher, Kim Sterelny, Daniel Stoljar and Suzanne Uniacke, and especially to Robert Kirby, Holly Lawford-Smith, R. J. Leland and Terry Macdonald for their written comments. Work on this chapter was supported by grants from the Australian Research Council and the Research Council of Norway.

[2] Although individual agents are discussed in this chapter, the analysis can be extended to collective agents as well.

of harm? Second, if there are such moral reasons, what is their basis?[3] Third, do overdeterminers have moral reasons to provide compensation or other forms of reparation to those who suffer overdetermined harms? Fourth, what should overdeterminers of harm do with the benefits that they have derived from their conduct? While all of the issues raised by these questions are important (and clearly interconnected), the first two – concerning reasons against overdetermining harm and their moral basis – seem most fundamental. If there were moral reasons against overdetermining harm, it would be relatively easy to understand why overdeterminers would owe compensation to those who have been harmed, and why they would be required to share the benefits that their overdetermining conduct produces.

It is not always obvious whether or not the conduct of particular agents overdetermines harms or whether, on the contrary, the conduct of each is necessary for the occurrence of some particular harm. Individual contributions to climate change may be like this. It is certainly *possible* that an individual person might, through their carbon emissions, make some difference to the occurrence of harm to other particular people, in the sense that if they acted differently these people would not have been harmed. If this were generally the case, then the issue of overdetermining harm would not be of critical importance to moral thinking about individual responsibility relating to climate change. But in general it seems very unlikely that individual carbon emissions really make a difference to the occurrence of any particular harm.[4] Some authors avoid the implication that individual carbon emission is a case of overdetermining harm. In a recent article, for example, John Nolt tries to undermine the idea that 'the harm caused by an individual's participation in a greenhouse-gas-intensive economy is negligible'.[5] He estimates the harmful impact of the 'average American's' carbon emissions by dividing the harm that will be generated by the total amount of carbon emitted by Americans as a group, and then dividing it by the number of Americans. He concludes that by this measure the average American's carbon emissions will be responsible for the deaths or suffering of two future people. To be sure, if each individual American had reason to believe that refraining from emitting carbon excessively might make

---

[3] Following Joseph Raz, let us understand a 'reason' as a consideration in favour of some act, 'which by itself is sufficient to necessitate a certain course of action, provided no other factors defeat it': J. Raz, *Practical Reason and Norms*, 2nd edition (Oxford: Clarendon Press, 1990), p. 19. Those who claim that there are overdetermination-based moral reasons against φ-ing are committed at a minimum to the view that this reason will be sufficient to necessitate their refraining from φ-ing, provided no other factors militate in favour of φ-ing. They may differ, of course, about just how weighty overdetermination-based reasons are, relative to other sorts of reasons.

[4] W. Sinnott-Armstrong, '"It's not *my* fault": global warming and individual moral obligations', in W. Sinnott-Armstrong and R. B. Howarth (eds.), *Perspectives on Climate Change: Science, Economics, Politics, Ethics, Advances in the Economics of Environmental Resources* (Amsterdam: Elsevier, 2005), vol. V, pp. 285–307.

[5] J. Nolt, 'How harmful are the average American's greenhouse gas emissions?', *Ethics, Policy and Environment*, 14/1 (2011), 3–10, p. 3.

the difference to whether two future people die prematurely or not, this would give them a very stringent reason not to do so. But while Nolt's accounting exercise may have some uses for moral assessment, his finding is consistent with its being the case that the total amount of harm caused by carbon emissions would be the same whether or not any particular American changed the amount of carbon they emitted. If this is so, then it would be true of each American that their carbon emissions make no difference to the occurrence of harm to particular people – each could vanish from the face of the Earth without the reduced emissions meaning that two future people are saved from premature death. And if it is true of each American that they would not make such a difference to particular people, then they would be overdeterminers of harm in our sense. This is not to say that individuals are not overdeterminers of each and every harm engendered by climate change. John Broome, for instance, reports that an average adult in a rich country emits around 800 tonnes of greenhouse gasses (GHGs) in their lifetime, and claims that it is very likely that their *cumulative* emissions will indeed make difference to the occurrence of harm.[6] Assuming that Broome is correct, there will nevertheless be many harms that individuals overdetermine through their conduct, additional to whatever harms are counterfactually dependent on their individual carbon emissions. Consequently, the stringency of reasons to refrain from emitting carbon or to compensate for their doing so will depend in some measure on the moral status of overdetermining harm.

So it is important to examine the issues of moral reasons against overdetermining harm (we'll refer to these as overdetermination-based constraints) and their bases, and that's the task we shall undertake in this chapter. We survey some proposed rationales for these constraints and note some of the criticisms to which they seem vulnerable. We then propose what we take to be a more promising alternative account.

Preliminaries

Doing harm

In exploring the idea of overdetermination-based constraints, we shall restrict our discussion here to cases in which the *type* of conduct in question does harm, rather than merely fails to prevent it or contributes to it in some other way (e.g., by enabling it to occur, or by facilitating its occurrence).[7] While the proper way to characterize the distinction between doing harm and merely allowing or

---

[6] J. Broome, *Climate Matters: Ethics in a Warming World* (New York: W. W. Norton and Company, 2012), pp. 74–8.
[7] The distinctions between doing, allowing, and enabling harm are discussed in detail in C. Barry and G. Øverland, 'The feasible alternatives thesis', *Politics, Philosophy and Economics*, 11/1 (2012), 97–119.

enabling it to occur remains controversial, clear-cut instances of doing harm typically have two features, and we will discuss cases in which the agent's conduct possesses them.

The first feature is what might be called *relevant action*. If Sue is linked to John's injuries by relevant action, then there is an answer to the question of *how* she was relevant to his injuries that refers to some act of hers. In a car crash in which Sue's car runs over John, Sue becomes relevant to John's broken leg *by driving into him*. The answer to the how question refers to an action of hers – her driving the car in a particular way and at a particular time. The second feature is that there is a *complete causal process* that links Sue's relevant action to John's injuries.[8] That is, there is an intact sequence linking the relevant action of Sue with the fracture of John's leg. In a car crash this intact sequence takes the form of a physical process involving the transfer of energy and momentum from Sue to John – a complete energy momentum sequence connects them.[9]

All other things being equal, moral reasons associated with doing harm that possess these features are commonly thought to have several important normative characteristics. First and foremost, there are stringent constraints against engaging in such conduct. They are stringent in the sense that prospective doers of harm cannot easily justify their conduct by appealing to the costs to themselves of refraining from doing harm, nor by appealing to the overall good that their conduct will bring about. And they are stringent in the sense that they demand much of agents who have ignored these constraints, but are now in a position to mitigate or alleviate the harm that they have done.

Sue has a stringent moral reason not to drive into and maim an innocent person, even if it is the only way she can avoid losing her own hand, or protect her child from suffering a broken leg. On the other hand, the fact that Sue would lose a hand or that her child would suffer a broken leg were she instead to intervene in a traffic incident to help an innocent person escape significant injury is ordinarily thought to provide justification (or at least a very good excuse) for her failure to help. In addition, the potential victims have claims against prospective doers of harm that they not harm them. Moreover, these claims are *enforceable* – potential victims (or third parties acting on their behalf) can enforce these claims through the proportional use of force. For

---

[8] Ned Hall (N. Hall, 'Non-locality on the cheap? A new problem for counterfactual analyses of causation', *Noûs*, 36/2 (2002), 276–94; N. Hall, 'Two concepts of causation', in J. Collins, N. Hall and L. A. Paul (eds.), *Causation and Counterfactuals* (Cambridge, MA: MIT Press, 2004), pp. 225–76) refers to causal connections of this kind as exhibiting 'locality'.

[9] This is not the only form that such processes can take. If John opens a dam and the water floods the town below, he does not *transfer* energy to the dam or the water, but rather *releases* stored energy that was being held at bay by the dam. For a short useful discussion of the different ways the idea of a complete process might be conceived, see J. Schaffer, 'Overdetermining causes', *Philosophical Studies*, 114 (2003), 23–45, p. 32.

instance, Sue may be prevented from killing a pedestrian with her car even if this involves injuring her significantly. Even when it is, all things considered, permissible or even obligatory to do harm to innocent non-threatening people (bystanders), compensation is typically owed to those who are harmed. This is because their stringent claims against having harm done to them have been infringed, however justifiably.[10]

All we have done so far is describe what we take to be widely accepted views about the normative significance of doing harm, and the way this category differs from allowing harm. Whether moral reasons associated with doing harm *really* have such significance remains a matter of philosophical debate. We won't join in this controversy here. We shall simply assume that reasons based on doing harm possess the characteristics that common sense morality accord to them, and plumb the significance of this assumption for the issue of overdetermining harm. Do particular overdeterminers harm others? The answer to this question depends very much on the conception of harm that is adopted. Some conceptions of harm that make counterfactual *dependence* essential for the attribution of harm will clearly not treat individual overdeterminers as harming others.[11] Conceptions of harm that make the attribution of harm depend on whether an individual is involved in the *production* of bad states of others, on the other hand, may count individual overdeterminers as harming.[12] Since the questions we have raised concerning the moral status of overdetermining harm are of interest whether or not individual overdeterminers are taken to harm others, we shall bypass discussion of the appropriate account of harm.

What is overdetermining harm?

To help fix ideas and make our discussion more concrete, let us introduce a few variations of a simple imaginary case that we hope can shed light on the more complex cases – such as individual carbon emissions – with which this volume is concerned.

A person – Robinson – is living relatively well on his small island. However, the water in the lake surrounding his island starts to rise. At some point the

---

[10] For discussion, see J. Thomson, 'Some ruminations on rights', *Arizona Law Review*, 19 (1977), 45–60.

[11] For discussion see M. Hanser, 'The metaphysics of harm', *Philosophy and Phenomenological Research*, 77/2 (2008), 421–50; J. Thomson, 'More on the metaphysics of harm', *Philosophy and Phenomenological Research*, 82 (2011), 436–58; and B. Bradley 'Doing away with harm', *Philosophy and Phenomenological Research* 85/2 (2012), 390–412.

[12] For discussion see S. Shiffrin, 'Wrongful life, procreative responsibility, and the significance of harm', *Legal Theory*, 5 (1999), 117–48; and E. Harman, 'Harming as causing harm', in M. A. Roberts and D. Wasserman (eds.), *Harming Future Persons* (Dordrecht: Springer, 2009), pp. 137–54.

island is swallowed up as a result of the rising water levels, and Robinson drowns. Fifty-one people have engaged in a type of conduct – shovelling excess waste from their gardens into the lake – that has led to the rising water levels. Tom is one of the fifty-one: he digs in his garden to extract some valuables, and shovels the excess waste into the lake. Tom's conduct alone would not have resulted in Robinson's death. Unfortunately, fifty other people also shovel their waste into the lake. As a result, water levels rise and Robinson is drowned.

Consider three versions of this story (only one of which, it turns out, is an instance of overdetermining harm, as we have defined this notion).

Robinson I: Tom's disposing of his waste is necessary for the flooding of the island and the drowning of Robinson. Although fifty others shovel waste into the lake, it is nevertheless the case that if Tom were to have abstained, Robinson would not have drowned.

Robinson II: Tom's disposing of his waste makes no apparent difference to the drowning of Robinson. Fifty others shovel waste into the lake and that number, or less than that number, is enough to drown Robinson. If Tom were to have abstained, the island would still have been flooded, and Robinson would still have drowned.

Robinson III: Tom's disposing of his waste makes no apparent difference to the drowning of Robinson. He shovels his waste into the lake after many others have disposed of their waste, and after this has already led to the flooding of the island and the drowning of Robinson.

We'll assume throughout this chapter (unless otherwise specified) that the situations are transparent to Tom and others in each of these cases. In Robinson I, Tom is *not* an overdeterminer with respect to Robinson's death. Why? Because Robinson's death would in this case be counterfactually dependent on Tom's conduct. Had Tom refrained from shovelling his waste into the lake, Robinson would not have drowned. In this scenario Robinson's vulnerability is obviously a very good reason for Tom not to shovel his waste into the lake.[13]

In Robinson III, Robinson's death is not counterfactually dependent on Tom's conduct. Moreover, it is clear that there is no intact causal sequence that links Tom's conduct with Robinson's death. Robinson has already drowned when Tom shovels his waste into the lake. Consequently, Tom is not an overdeterminer of Robinson's death in this case. In the absence of some independent and negative consequence of his disposing of his waste in this way (perhaps it shows disrespect of some sort), it is hard to see why Tom has any reason to refrain from acting as he does in Robinson III. It does not seem that there is any claim of Robinson's that would be infringed by Tom's conduct.[14]

---

[13] This is precisely the sort of reason Broome stresses when condemning carbon emissions without counterveiling offsets (see Broome, *Climate Matters*, esp. pp. 50–5, 96).

[14] Insofar as compensation is owed to Robinson's relatives, it would seem to be owed by the others who shovel waste prior to the water reaching the stage where it causes his death, but not by Tom.

In Robinson II, by contrast, Tom is an overdeterminer of Robinson's death. Robinson's death is not counterfactually dependent on Tom's conduct (he would have died whether or not Tom disposed his waste), but it is nevertheless the case that if Tom and enough others shovel their waste into the lake, Robinson will drown. Further, there may be an intact sequence linking Tom's conduct and Robinson's death in Robinson II (but, since it is overdetermined, there may not). If there are overdetermination-based constraints, then all else being equal it would be wrong of Tom to dispose of his waste in Robinson II.

Existing approaches to overdetermining harm

Scepticism

It might be argued that there is no constraint against Tom's disposing of his waste in Robinson II. After all, one might reasonably believe that the reason why there are constraints against certain types of conduct is that by engaging in them one tends to make other people worse off, or at least imposes a risk on others that they might be made worse off. In situations of overdetermination, however, this is not the case. Robinson will not be made worse off if Tom shovels his waste into the lake on this particular occasion. Recall that we are assuming that the fact that the conduct will make Tom an overdeterminer is transparent to him and others. The sceptic claims that while it might be true that the agent's particular action would ordinarily have been wrong because it would have done harm to the victim, he does nothing wrong by performing that action in this particular case, because he makes no difference to what happens to the victim. There is some evidence that the law, at least, frees overdeterminers of harm from responsibility in some cases.[15] This line seems consonant with the general view of morality endorsed by Frank Jackson:

> the morality of an action depends on the difference it makes; it depends, that is, on the relationship between what would be the case were the act performed and what would be the case were the act not performed.[16]

This view has been sympathetically explored (though not endorsed) in application to the case of climate change by Walter Sinnott-Armstrong.[17]

---

[15] These are the asymmetrical concurrent cause cases, the symmetrical concurrent cause cases where one sufficient cause is a natural event, and the damage limitation rule in cases of preemptive causation. For discussion, see M. Moore, 'For what must we pay? Causation and counterfactual baselines', *San Diego Law Review*, 40 (2003), 1181–271, pp. 1264–66.
[16] F. Jackson, 'Which effects?', in J. Dancy (ed.), *Reading Parfit* (Oxford: Blackwell, 1997) pp. 42–53.
[17] Sinnot-Armstrong, '"It's not *my* fault"'. See also the discussion in Garrett Cullity's chapter in this volume.

The sceptic claims that there are no overdetermination-based constraints. He believes that all else being equal Tom has no reason not to dispose of his waste in Robinson II. When Tom shovels his waste into the lake, he obtains benefits without worsening anyone's situation, and this gives him a good reason to do it. Correspondingly, Robinson has no claim against Tom that he refrain from the overdetermining conduct. Of course, many people seem to think that there *is* something wrong about Tom's conduct in Robinson II. They find it problematic that Tom can φ when φ-ing is a type of action that alone or in conjunction with the actions of others could bring about severe harm, even though on this particular occasion the harm will occur whether or not Tom φs, because many others will φ. But the sceptic may be able to provide plausible explanations for this intuition without attributing any moral significance to conduct that overdetermines harm. For example, they can argue (in a Lockean vein) that because we strongly associate φ-ing with certain harms, this leaves a kind of psychological trace that we cannot easily free ourselves from. This inclines us to condemn an agent's φ-ing even when her φ-ing is not necessary for the harm to occur.

Still, scepticism about overdetermining harm is a powerful position, and one that we shall not attempt to disprove here. The appeal of scepticism could be blunted considerably if a plausible positive account of these constraints can be provided, however, and that is what we shall try to do.

So let's explore some existing justifications for overdetermination-based constraints to see whether they hold promise.[18] Our aim in exploring them is not to provide knockdown arguments against them. Rather, we indicate the sort of objections to which they are vulnerable. We think these objections are serious (even if they could ultimately be met) and warrant exploration of alternative justifications for constraints against overdetermining harm that are not vulnerable to them.

Absolutism

It might be argued that since φing is the sort of action that typically has bad consequences, it is simply wrong to φ, even if in certain circumstances φing makes no-one worse off. Following Jonathan Bennett, let us distinguish between *absolutism* and *relativism* about types of conduct.[19] Absolutism

---

[18] Our list includes what we take to be the most promising existing justifications for the constraints in question, and is not exhaustive (see Sinnott-Armstrong, '"It's not *my* fault"' for some other arguments that we do not consider).

[19] As Bennett (J. Bennett, *The Act Itself* (New York: Oxford University Press, 1995), pp. 165–71) notes, when contrasting absolutism and relativism we ought not to describe the kind of behaviour whose moral status is in question in terms of its *overall* consequences. There is substantive dispute between absolutism and relativism about the wrongness of engaging in some conduct when that conduct is described as 'killing an innocent person when there is no good to be achieved by so doing'. The conduct must be described in such a way as not to invoke an overall moral

about φing is the thesis that it could *never* be right to φ. Relativism about φing is the thesis that it can be permissible to φ in some possible circumstances, but that it may not be permissible in others.

It is controversial whether absolutism about any type of conduct is plausible.[20] However, even if one could make a case for absolutism about some types of conduct – shooting or torturing innocent non-threatening people, for example – it is extremely unlikely that a case can be made for absolutism about many of the types of conduct that are involved in overdetermination cases. If no one were made worse off or rendered unduly vulnerable as a result of our emitting high levels of carbon or shovelling waste into lakes, how could it possibly be wrong for us to do these things?

The universalisation requirement

One might appeal to Kant-inspired arguments as a basis for overdetermination-based moral reasons. That is, in deliberating about whether or not they can φ, the agent must not think in terms of whether or not their φing would make a morally relevant difference in this instance, but whether it would make a morally relevant difference if *everyone* φed, or (to take another version of the universalisation test) if everyone who was *inclined to* φ did so.[21] Clearly, the conduct that universalisation tests would rule out as impermissible would seem to depend very much on how the conduct is described. If Tom's conduct in Robinson II is described as 'shovelling waste into the lake', then perhaps this test will require that the agent not engage in it, since morally unacceptable consequences (e.g., Robinson's death) would result were everyone to do it, or even if everyone who is inclined to do so does it. If the conduct in Robinson II is instead described in terms of 'shovelling waste into the lake when doing so makes no difference to anyone', then it is likely that the conduct would not be ruled out by the universalisation test. As Parfit puts it, 'if my contribution would make no difference, I can rationally will that everyone else does what I do. I can rationally will that no one contributes when he knows that his contribution would make no difference.'[22]

One could insist that in formulating the universalisation test we should describe the conduct in a way that does not include the incorporation of any contextual features based on what others will do or are likely to do. Allowing

---

assessment – it is not controversial that there is an absolute moral prohibition against 'impermissibly killing someone'.

[20] G. E. M. Anscombe, 'War and murder', in W. Stein (ed.), *Nuclear Weapons: a Catholic Response* (London: Burns and Oates, 1961) pp. 43–62; Bennett, *The Act Itself.*

[21] See T. Pogge, 'The categorical imperative', in P. Guyer (ed.), *Kant's "Groundwork of the Metaphysics of Morals": Critical Essays* (Lanham, MD: Rowman and Littlefield, 1998), pp. 189–213 for discussion of this version of the test.

[22] D. Parfit, *Reasons and Persons* (Oxford: Clarendon Press, 1983), p. 87.

the incorporation of such features seems to invite agents to concoct 'gimmicky' maxims that effectively *circumvent* the universalisation tests, rather than apply them sincerely.[23] The main drawback to this proposal is that, if we exclude such contextual information, the universalisation test is very likely to yield results about how agents ought to behave that in many cases are difficult to accept. It seems very important that we be able to consider the likelihood that other people will behave in certain ways when we are deliberating about what we are permitted to do, rather than assuming away such information. To borrow an example of Jonathan Glover's, I may not be able to rationally will that everyone be heavily armed, but it does not follow from this that I may not heavily arm myself when I know that others around me are heavily armed.[24]

Overdetermination as contribution: necessary elements of a sufficient set

A very different proposal for justifying overdetermination-based constraints is to say that overdeterminers contribute to overdetermined harms, not by becoming a difference maker with respect to their occurrence, but by becoming necessary elements in a set of actual antecedent conditions that is sufficient for bringing those harms about – a 'NESS' condition for their occurrence.[25] An agent's conduct is a NESS for some harm only if it was a necessary element of a set of antecedent actual conditions that was sufficient for its occurrence. Richard Wright explains this notion as follows:

[A] particular condition was a cause of (condition contributing to) a specific consequence if and only if it was a **N**ecessary **E**lement of a **S**et of antecedent actual conditions that was **S**ufficient for the occurrence of the consequence.[26]

According to some theorists, NESS conditions provide the best available analysis of the concept of causation.[27] Wright, for example, claims that the NESS idea 'not only resolves but also clarifies and illuminates the causal issues in the problematic causation cases that have plagued tort scholars for centuries', but that it 'is the essence of the concept of causation'.[28]

---

[23] Pogge, 'The categorical imperative'.
[24] J. Glover, *Causing Death and Saving Lives* (Harmondsworth: Penguin, 1977), p. 131.
[25] This view of contribution is developed by Hart and Honoré (H. L. A. Hart and T. Honoré, *Causation in the Law*, 2nd edition (Oxford: Clarendon Press, 1985)). They acknowledge that it builds on J. S. Mill's notion of a jointly sufficient set of conditions, as well as on Mackie's idea (J. L. Mackie, *Cement of the Universe*, 2nd edition (Oxford University Press, 1980)), in the context of causal generalisations, of an INUS condition (an Insufficient but Necessary part of an Unnecessary but Sufficient condition).
[26] R. W. Wright, 'Causation in tort law', *California Law Review*, 73 (1985), 1788–813. Capitals added to clarify the acronym.
[27] Hart and Honoré do not make any such claim.
[28] Wright, 'Causation in tort law', pp. 1802 and 1805.

In Robinson II, Tom's conduct is indeed a necessary element of some actual set of conditions that was sufficient for its occurrence, even if disposing of the waste of forty would suffice to flood the island. When Tom throws waste into the lake in this case, his conduct becomes a necessary element in *an* actual set of conditions comprising it and, for instance, the conduct of others (1, 2, 3 . . . 38 and 39) who shovel waste in the lake. Together these forty people would flood the island. And this, of course, is true of each of the fifty-one waste throwers. It is true of each that their waste disposal was a necessary element of *a* set of actual antecedent conditions – namely, any set excluding the waste disposal of eleven others, but including that of thirty-nine others, which was sufficient to bring about the outcome. A plurality of sets of actual antecedent conditions was sufficient for Robinson's death. It is not the case that the conduct of any one of the fifty-one waste throwers was necessary for the sufficiency of all of these sets, but the conduct of each was necessary for the sufficiency of some of them.

On this view, there are overdetermination-based constraints against Tom's throwing waste in Robinson II, because by doing so he would become a contributor (NESS) to the outcome. Insofar as we accept that the reasons against becoming a contributor to harm are ordinarily quite stringent, then there is nothing mysterious about there being such reasons against Tom's shovelling his waste into the lake in Robinson II.

While this approach might rescue the idea that there are overdetermination-based constraints, in our view it does so by invoking an extremely implausible notion of contributing to harm. A well-known example from the literature on causation helps illustrate this.[29] Bill and Ben each throw a rock towards a window. The window shatters. They are convicted of vandalism, and are required to compensate the shopkeeper. Alice has it all on tape. A closer look reveals that it was only Bill's rock that hit the window. Ben's rock followed its trajectory soon thereafter, passing through the shattered glass, but without hitting it – it did not change the timing or the manner of the occurrence of the shattering of the window. To say that Ben contributed to the shattering of the window when neither he nor his rock came into contact with the glass (and where no energy was transferred from the approaching rock to the glass) is implausible. Ben was accordingly not a contributor to the breaking of the window, and any understanding of contribution that maintains that he did contribute to this outcome is implausible.

Some legal theorists nevertheless seem to favour this NESS account of contribution in allocating costs for accidents and punishments for crime.[30] It is not

---

[29] Cf. D. Lewis, 'Causation as influence', *Journal of Philosophy*, 97/4 (2000), 182–97.
[30] J. Stapleton, 'Choosing what we mean by "causation" in law', *Missouri Law Review*, 73 (2008), 433–80.

difficult to understand why, given that many legal systems currently make causal contribution to harm a condition for liability to bear cost to address it.[31] One might be concerned that overdeterminers will be let off the legal hook unless one adopts a NESS account of causal contribution, or something akin to it. The correct response to this, however, is not to pretend that an agent contributes to something by saying that he is a NESS of it. Ben does not contribute to (although he is a NESS of) the breaking of the window in the case imagined above. If we believe that overdeterminers should be made liable for the costs of their conduct, we should instead conclude that there are good grounds to change the requirements for liability. There might be good reasons to make both Bill and Ben bear the cost of repairing the window, and good reasons perhaps to punish them. Making actual contribution to harm a requirement for duties to provide compensation to the victims of harm could be questionable, morally speaking.[32]

An alternative approach

Levels of description

Does overdetermination even exist? Some simply deny the existence of overdetermination in a world like ours, where there are certain lawful physical processes.[33] To be sure, there are many situations in which some set of people *appears* to consist of overdeterminers with respect to some injury, but perhaps this is only because we lack the necessary detailed microanalysis of the situation and how it has come about. Accordingly, the question 'what is wrong with being an overdeterminer of harm?' would not even get off the ground.

Harms and other outcomes can be described generally – a window is broken, Robinson dies – or they can be described more specifically – the window breaks and Robinson dies at particular times and in particular ways. One can clearly be an overdeterminer of harms that are described at a very general, rough-grained level. However, it is very often the case that a person is not an overdeterminer with respect to the outcome when it is described in a more fine-grained fashion. Whether a person is an overdeterminer with respect to some outcome, then,

---

[31] American Law Institute, *Restatement (Second) of Torts*. §§431–433 (1965).
[32] J. Waldron, 'Moments of carelessness and massive loss', in D. G. Owen (ed.), *The Philosophical Foundations of Tort Law* (Oxford: Clarendon Press, 1995) pp. 387–40. Shelly Kagan has recently tried to address our question regarding overdetermination-based constraints from a consequentialist perspective. We do not think his arguments are successful (and try to show this in forthcoming work) but cannot in engage with his views here due to constraints of space. S. Kagan, 'Do I make a difference?', *Philosophy and Public Affairs*, 39 (2011), 105–41, pp. 117–18.
[33] See M. Bunzl, 'Causal overdetermination', *Journal of Philosophy*, 76/3 (1979), 134–50. For arguments against this view, see Schaffer, 'Overdetermining causes', who claims that overdetermination is 'everywhere'.

seems to depend on the level of detail at which that outcome is described. It may well be that in a world such as ours no outcomes are overdetermined when they are described at a sufficiently fine-grained level of detail.

A proposal: probability of being a member of the actual set

A fine-grained analysis seems to us to provide the correct diagnosis of who contributed to Robinson's death. It fails, however, to explain the moral reasons that apply to Tom when he is considering whether he can throw waste into the lake. Learning whether or not Tom actually did or did not contribute to Robinson's death as described in this fine-grained way doesn't really explain why it is wrong for him to dispose of his waste in the first place.

The proposal here is that overdetermination-based constraints are based on the possibility that some agent will become a necessary element of *the* set of actual conditions that *actually* brings about the overdetermined outcome. The significance of the overdetermination-based constraint is therefore in an important way a function of the probability that by φing an agent will be a member of *this* set. If the agent φs, then an overall moral appraisal of his conduct would also depend on the costs to him of refraining from φing and on his knowledge of the situation that he is in, as well as the benefits his conduct may produce for others. The higher the costs to him, the less transparent the situation, and the larger the benefits his conduct will produce for others, the less culpable he will be.

This account of overdetermination-based constraints can be summed up as follows:

The significance of overdetermination-based constraints (against φing) is a function of how bad the overdetermined outcome is and the probability that by φing this agent will be an element of the set of actual antecedent conditions that brings this outcome about.

Assuming that the badness of the outcome is held constant, then the more unlikely it is that he will be among the actual set by his φing, the less significant his constraint against φing will be. If there is no risk that by φing the agent will be among the actual set of conditions that in fact does the harm, then there are no overdetermination-based constraints against her φing.[34]

Consider how our proposed account applies to variations of Robinson II. Assuming that the number of waste throwers necessary to cause him to drown is forty, then the significance of Tom's overdetermination-based constraints against throwing waste into the lake is reduced when the number of other people throwing waste into the lake increases. This is because when the number of waste throwers increases, the probability that Tom will be a member of the

---

[34] The conduct may, of course, be wrong for some other reason.

actual set decreases. Holding the necessary number of waste throwers constant at forty his probability will be 40/40, 40/41, 40/42, and so on.

When holding the number of people throwing waste constant at fifty-one, the significance of Tom's overdetermination-based constraint against throwing waste into the lake will increase as the number of waste throwers required to flood the island increases. This is because the larger the number of waste throwers necessary to bring about Robinson's drowning, the greater the likelihood that Tom will be a member of this set. For instance, if only one person is required, then the probability that he is a member of the actual set is 1/51, if two people are required, then the chance that he is among them is 2/51, and so forth, until the required number is fifty-one, in which case he will certainly be a member of this set.

Is the number of contributors relevant?

When only one person needs to φ to bring about an outcome, and there are fifty-one people who φ, then the probability that any particular individual will be a member of the actual set is much lower than if a larger number of people were required to bring it about. According to our proposal, this means that the overdetermination-based constraint against throwing waste will be correspondingly weaker on each than if a larger number of people were required. But perhaps there is an additional factor that we have failed to consider that might militate against this conclusion: the absolute number of people whose conduct will in fact bring about the outcome.[35] If Tom should happen to be a member of the actual set that brought about the flooding when only one person is required to flood the island, he would be the lone contributor to Robinson's death. This might seem to make a difference. Perhaps the size of the actual set is independently relevant for determining the significance of an agent's constraint against becoming an overdeterminer.

If this were true, then the size of the actual set necessary to do harm would be relevant for two separate assessments that together would determine the significance of an agent's overdetermination-based constraint: (1) the probability that any given person will be a member of the actual set is relevant for gauging the significance of the overdetermination-based constraint against φing (the smaller the set, the less likely they are in it, and the less wrong their φing), and (2) the amount of harm that each person in that set is responsible for is relevant for the general wrongness of being a member of the actual set (the smaller the set, the more harm each member of the set is responsible for).

---

[35] Thanks to Geoff Brennan for pointing out this possibility.

These two assessments could pull in different directions. The probability that an agent will become a member of the actual set by φing could be low. Correspondingly, the overdetermination-based constraint against φing is weak (on the first dimension). But if the agent nevertheless ends up as a member of the actual set, his role in bringing it about is substantial – he might even be the only one in the set – in which case his overdetermination-based constraint against φing would be stringent (on the second dimension.) Whether the size of the actual set necessary to do harm matters is of particular importance in the case of individual contributions to climate change. The size of the actual set needed to bring about harmful climatic processes is no doubt *very* large. Thus, the 'share' that any individual will have in any particular harm would seem miniscule. So if the size of the set matters, this will correspondingly mean that individuals will have much less reason to refrain from emitting excessively than they would, given the severity of the harms that will result from the climatic processes to which they contribute.

However, we are not convinced that the size of the set matters. Why should it matter that you are one of many when you are an element in the actual set that brings the harm about? Consider Robinson II. If it takes fifty-one people to kill Robinson, and that is well known by all, it is not clear that any one of them is any less to blame for his death than if one of them were to drown him alone. After all, each of them is in a position to avoid killing Robinson by refraining from pushing waste into the lake. That one person's contribution depends on other factors to be effective doesn't seem to alter the extent of their responsibility when they know that these factors obtain, regardless of how many additional necessary factors there are. Looked at from another angle, each of these contributors would, it would seem, be liable to defensive force (should it become possible to defend Robinson) that would be comparable to that of one person whose conduct alone would be sufficient to drown him. As noted above, this seems at least a good indicator of the cost they would be obliged to take on to avoid contributing to harming Robinson.

Nor is it obvious that the size of the actual set is of independent relevance, even when the agents involved do not know whether the other necessary factors obtain. When Tom considers whether or not to shovel waste into the lake, his uncertainty about whether or not enough others have done so that his conduct would combine with theirs to bring about Robinson's death is obviously relevant to his decision. But it is not clear whether the *numbers* of factors that would have to be present for the outcome to be brought about is relevant in itself. When the conduct of many people is necessary for bringing about an outcome, then it will often be more difficult for any particular person to ascertain that they will become a contributor to this outcome if they φ. The true significance of numbers could therefore simply be that the more factors that are necessary to bring about some outcome, the more likely an agent is to be uncertain about whether his conduct will be necessary for it to occur. This seems of particular relevance to the case of climate change, where the scale and complexity of the system make it very unclear how the conduct of any individual will play a role in bringing about any particular climate-related harm.

Stringency

Our focus so far has been on proposing a way of understanding overdetermination-based constraints that undermines the appeal of scepticism about reasons against overdetermining harm. But what of the view that Tom's being an overdeterminer (rather than a necessary condition) of Robinson's drowning makes no difference to the moral reasons that apply to Tom?[36] That is, the constraints that apply to him are just as stringent as those that would apply were his conduct instead to be a necessary condition for the harm.

This position is implausible. The fact that the bad outcome will happen whether or not Tom φs clearly should make a difference to the weight of reasons against his φing. This is supported by considered judgements in a range of cases. It seems to make a significant moral difference whether one is exposing someone to 1/51 risk of death or is instead an overdeterminer with a 1/51 risk of being among the actual set that kills a person. In both cases there is some possibility that you will kill him. But only in the former case is there a possibility that you will make a difference to whether or not he is killed. There seems also to be significant moral difference between Tom's pulling the trigger of a Russian roulette gun with fifty-one chambers and his being one of fifty-one people shovelling waste into a lake (or one of fifty-one shooting at a person's head, for that matter), when it takes only one to kill the victim.

The fact that the bad outcome is going to happen whether or not one φs seems to function as a factor that reduces the stringency of moral reasons against φing. Moreover, the degree to which it reduces the significance of the overdetermination-based constraint depends crucially on the motives of the agent who is φing, and the interests that are served when he φs. As originally described, Tom shovels waste into the lake in Robinson II simply to get rid of it, because he can reap benefits from doing so, and for no other reason. However, we could modify the case to allow him other reasons for throwing the waste into

---

[36] This seems to be the position defended in A. I. Goldman, 'Why citizens should vote: a causal responsibility approach', *Social Philosophy and Policy*, 2 (1999), 201–17; and R. Tuck, *Free Riding* (Cambridge, MA: Harvard University Press, 2009), pp. 32–60.

the lake. For instance, he might need to get rid of it to save another person from drowning. In that case, it seems that the reasons against his throwing waste into the lake are outweighed by his positive reasons to save that person. And that seems to be the case even when the probability that he will be a member of the set that actually does harm to Robinson is very high. By contrast, it does not seem that he would be permitted to φ in order to save another person if his φing would drown Robinson when it is *not* the case that Robinson would have drowned anyway – we cannot do harm to one innocent person to save one other innocent person from a comparable harm to which we would not be a contributor.

Earlier in this essay we noted that it was widely accepted that doers of harm cannot easily justify their conduct by appealing to the costs to them of refraining from doing harm, or to comparable harms that their conduct will prevent for others if they do harm. Our claim here is that it is much easier for overdeterminers to appeal to costs of these kinds to justify their conduct. Overdetermination-based constraints are simply not as significant as constraints against doing harm, when your conduct is necessary for harm being done. That the bad outcome is going to happen increases the plausibility of the appeal to cost or to prospective benefits for others.

Could Tom invoke just *any* reason to protect another person from harm to override his overdetermination-based constraint against throwing waste when the island will be flooded whether or not he does so? If he could, this constraint would be extremely weak – hardly worth its name. It may seem plausible that these constraints are very weak when the probability that Tom will be a member of the actual set is very low. On the other hand, when the probability that he will be among the actual set is high, the fact that his conduct would bring minor benefits to him or others does not seem sufficient to override this constraint. If Alice needs to tidy up her garden, it seems wrong of Tom to help her by throwing the waste into the lake if the situation is as described in Robinson II. This is so even if the probability of his being among the actual set that causes the flooding is quite low – if fifteen rather than forty would suffice to bring it about. Although Robinson is going to drown anyway, Tom shouldn't throw waste into the lake unless an interest of considerable moral significance is served by his doing so. And it seems that if he throws waste into the lake without this sort of justification, he would become liable to defensive force by Robinson or others acting on his behalf.

What if Tom were to throw waste into the lake in Robinson II in order to benefit himself? It seems pretty obvious that he could do so if it were necessary to save his own life or avoid a very significant loss. But what if he instead stands to gain only significant monetary benefits? Suppose he is not throwing the waste simply to get rid of it, but because he must dispose of it in order to extract some precious stones on his property. Suppose also that the only way to get rid of the waste is to throw it into the lake. Similar considerations apply in the case of individual emissions. Insofar as individuals are overdeterminers of harms resulting from climate change, the purposes to which their emissions are put seem to matter a great deal. Surely, emissions that are related to ensuring subsistence contribute just as much as any other kind of emission to climate change, but as Henry Shue has pointed out, their status seems different, given what is at stake for the agents involved.[37] This is what makes so-called 'luxury emissions' seem objectionable, even when individuals responsible for them are merely overdeterminers of harm.[38]

As noted initially, the issue of what to do with the *benefits* of overdetermining conduct is a distinct matter that we cannot address in detail in this chapter. However, it seems to us that Tom should not proceed in this case and claim all the benefits for himself in Robinson II. Even when the probability that an agent is a member of the actual set that brings about some harm by φing is very low, he seems required to share the benefits with those harmed when this is possible, or with relevant third parties (e.g., relatives) when it is not. When the overdeterminer φs, he violates a constraint and correspondingly infringes the claim of the person that suffers the overdetermined harm. That he infringes a claim explains why it is impermissible for him to keep all the benefits for himself. If Tom ends up in the actual set that brings about the outcome (described in a fine-grained way), then he seems to have no claim on the benefits he gains at all, and ought to compensate the victim as well. This too is relevant in the case of climate change. Insofar as people reap benefits from overdetermining harm, they acquire a duty to those whose harms they overdetermine. The stringency of this duty will depend, as noted above, on how essential the benefits are to the normal functioning of the individual, and whether they are emitting within their 'fair share'. If an individual emits only so much as is needed to meet their basic needs, or only so much as would be consistent with all others emitting as much without harmful consequences, then their duties to redistribute the benefits of conduct that overdetermines harm would seem to be sharply reduced, if not eliminated. If on the other hand individuals emit far in excess of these norms, their duties to redistribute the benefits to those harmed are quite stringent indeed.

Compensation

It is not obvious why anyone should provide compensation for victims of some harmful activity unless their conduct was in some way actually

---

[37] See H. Shue, 'Subsistence emissions and luxury emissions', *Law and Policy*, 15/1 (1993), 39–59.
[38] This is not to say that it is easy to distinguish between these types of emissions, though there will be instances of emissions that seem to fall clearly into each. See D. Jamieson, *Reason in a Dark Time* (New York: Oxford University Press, 2014), pp. 155–8 for discussion.

implicated in the harms that they have suffered. Compensation is due, in the first instance, from those who ended up among the set that actually brought about the bad outcome, described in a fine-grained way. Sometimes, of course, we will be unable to determine who actually brought about the outcome described in such a way. In these cases, we propose that all of those who engaged in the activity ought to share equally the cost of compensation, all else being equal. Things would not be equal if the probabilities that different agents are elements in the actual set differ, or if some of these agents are more culpable than others. In this case those with higher probabilities of being in the actual set or who are more culpable would owe correspondingly more. Thus, those who emit far in excess of their 'fair share' would owe correspondingly more than those who do not. How is it that we can require these people to compensate when we cannot tell for sure that they actually contributed to the outcome? We are entitled to do that, because these agents have all engaged in taking a risk of becoming an element in the actual set that brought about the outcome. It is therefore fair that they compensate the victim, since the overdeterminers are owed no benefit of the doubt.[39]

---

[39] We do not mean to deny that those who are not members of the set who actually bring about the harm may also be liable to provide compensation, but we are not endorsing this view either.

10 Climate change: life and death

John Broome

Ethics and danger

The United Nations Framework Convention on Climate Change (UNFCCC) is the treaty in which nations agreed to try and control climate change. It has been signed by virtually every nation on Earth. It declares in Article 2 that:

> The ultimate objective of this Convention ... is to achieve ... stabilization of greenhouse gas concentrations in the atmosphere at a level that would prevent dangerous anthropogenic interference with the climate system.

This article of the UNFCCC has set the agenda for subsequent political discussion and negotiations. To speak very roughly, the international process under the UNFCCC has three stages. First of all, the negotiators and their advisers try to work out what would be dangerous anthropogenic interference with the climate system. Then they try to work out how much greenhouse gas can be emitted without reaching this dangerous level they have fixed on. Finally, they try to reach an agreement about how to divide up those permissible emissions among the nations.

It is widely recognised that the last stage raises ethical issues. The allocation of emissions permits raises some difficult ethical questions. How far are the present members of a nation responsible for the harm that is done by their ancestors' emissions? How far are they responsible for the harm done by their own emissions made before they knew they were harmful? How far should the distribution of emission permits take into account the present maldistribution of so many of the Earth's other resources? Should it try to correct for this maldistribution? Should nations whose population grows be awarded increased emissions as a result? And so on.

It is widely recognised that there is a role for moral and political philosophers in answering questions like these. They are mostly questions of fairness or justice. Most philosophers of climate change are political philosophers, and political philosophy is these days very much focused on justice. So many philosophers of climate change are content to concentrate on this third stage of the political process.

A consequence is that the other two stages have been left mostly to scientists and economists. The question of what concentration of greenhouse gas would be dangerous is left mostly to them. Stemming from their work, a consensus has emerged in the political process around the ill-founded idea that it would be dangerous to let temperatures get higher than two degrees above the pre-industrial level.[1] In 2013, Working Group 1 of the Intergovernmental Panel on Climate Change (IPCC), which contains no philosophers, implicitly endorsed a target of keeping cumulative emissions below one trillion tonnes of carbon dioxide.[2] Their grounds are that this would give us a two-thirds chance of holding warming below the two degree level.

These are weak grounds. There is evidently an important role for moral philosophers in answering the question of what is dangerous. It will be the subject of this chapter. The notion of dangerousness is plainly an evaluative one. To work out what interference with the climate system is dangerous, we need to know, not only what effects would result from different degrees of interference, but also how good or bad those effects would be. Ethics, and specifically value theory, is the discipline that assesses goodness and badness. So we need ethics.

It seems not to be clearly understood by all the protagonists that there is an academic discipline that deals with values. The *Synthesis Report* of the Third Assessment Report of the IPCC – again written without a contribution from philosophy – declares that:

Natural, technical, and social sciences can provide essential information and evidence needed for decisions on what constitutes 'dangerous anthropogenic interference with the climate system'. At the same time, such decisions are value judgements determined through socio-political processes, taking into account considerations such as development, equity and sustainability, as well as uncertainties and risk.[3]

Here the authors recognise that dangerousness is not a matter for science alone, because it involves values. But they seem to assume that judgements about value have to be 'determined through socio-political processes'. They seem not to recognise that values can be investigated by philosophy.

True, in some cases the ethical aspects of dangerousness are so easy to understand that they raise no questions for philosophy. These are cases where

---

[1] The history of this consensus in described in C. Jaeger and J. Jaeger, 'Three views of two degrees', *Regional Environmental Change*, 11 (2011), 815–26. An example of target-setting by scientists is J. Hansen *et al.*, 'Target atmospheric CO_2: where should humanity aim?', *Open Atmospheric Science Journal*, 2 (2008), 217–31.

[2] The IPCC is officially 'policy-neutral'. However, the two-degree target and the trillion-tonne target are given prominence in Working Group III's *Summary for Policymakers*, 25. The trillion-tonne target was also evidently made prominent in the press conference that launched Working Group III's report, since it was widely reported in the press.

[3] IPCC, *Climate Change 2001: Synthesis Report,* question 1, section 1.1.

the physical conditions are like a cliff. It is not difficult to work out that you should not walk over a cliff, even if it is hard work or very costly to avoid doing so. The consequences of doing so are so obviously dire that you do not have to think about precisely how bad they are. All you need to know is where the cliff is. A map of the physical geography will tell you that. You need no advice about value.

Climate change could be like a cliff. It could be that the consequences of climate change will be only moderately bad so long as the concentration of greenhouse gas stays below some threshold, but disastrous if it goes beyond the threshold. For example, it might be that, if we pass some threshold, all the methane buried under the Arctic Ocean will bubble up and cause runaway global warming. This threshold would be like a cliff. If there is a cliff, and if scientists could work out where it is, there would be no need for moral philosophy. We would need only scientists to tell us where the cliff is. Scientists could determine the point where our interference with the climate system would become dangerous.

But even if there is a cliff, scientists have not been able to work out where it is. We do not know what concentration of greenhouse gas, if any, will take us over a cliff. One thing we do know is that the greater the concentration of greenhouse gas, the greater our risk of falling over a cliff. Suppose we assume that falling over a cliff would lead to some extremely large amount of harm. What is the expectation of that harm? I mean the mathematical expectation, which is the amount of harm multiplied by the probability of its happening. For a low concentration, the probability that we fall over a cliff is small, and so is the expectation of harm, therefore. With increasing concentration, the probability of falling over a cliff increases, and so does the expectation of harm. So, even if there is a cliff in terms of actual harm, because we do not know where it is, in terms of expectation we have a downward slope rather than a cliff.

There are also actual harms that increase with concentration in a way that is not cliff-like. For example, as temperatures and sea levels rise, farmland and people's homes vanish underwater, and farming becomes harder. These harms steepen the downward slope of expectations that is created by the risk of a cliff. Just as there is uncertainty about the cliff, there is a great deal of uncertainty about these harms too. We do not know how quickly temperatures and sea levels will rise. So expectations of harm are all we have to go on, and we face a downward slope of expectations as concentrations of greenhouse gas rise.

Given that, what concentration leads to dangerous interference with the climate system? This question needs to be framed more precisely. Some dangers are worth accepting for the sake of the benefits that can be gained by accepting them. For example, some dangerous surgery is worth the risk it poses. Evidently, the UNFCCC does not mean to refer to acceptable danger, but to danger so great that it ought to be avoided. Our question should therefore be: what concentration of greenhouse gas is so dangerous that we ought to keep below it?

Indeed, dangerousness is not really the issue. A situation is dangerous when there is a risk of a great harm – a risk of going over a cliff as I put it. Climate change poses that risk. But it also creates harms that increase steadily with the concentration of greenhouse gas. These are bad but not exactly dangerous. Nevertheless, they are an important part of the problem. So what we really need to know is simply what concentration we should keep below.

The 'should' in this question is a moral one. Science and economics can provide data to help answer it, but they cannot provide the whole answer. That also demands judgements of value. We can act more or less strongly to reduce emissions of greenhouse gas. Stronger action has the benefit of reducing concentrations, which reduces the expectation of harm. But stronger action has costs. For example, we could ban all air travel, and the eating of meat, and that would damage the interests of many people. It is not obvious whether or not the benefit gained would be worth this cost.

To know what actions are worthwhile, we have to balance costs against benefits. That requires setting a value on them. We have to assess how good and bad are the various different effects. Science can tell us what these effects might be, and how likely each one is, but it cannot tell us how good or bad they are. The necessary judgements of value ultimately have to be founded on ethics.

In practice, the detailed calculations of value will have to be done by applying the quantitative methods of economics, using data derived from science. This is just because climate change is a hugely complicated problem on a huge scale, involving the balancing and aggregating of different values across the whole globe and across centuries of time. But the foundations of these economic methods must be ethical. The economics of climate change is at heart an application of ethics.[4]

Expected value

Why do I say it is expectations of costs and benefits that we should care about? This claim is already a conclusion of ethical theory. It is by no means obviously true, although the idea of basing decisions on expectations is intuitively attractive once you see it. Since it will be important later, I shall introduce it by means of an example.

---

[4] As is well recognised in *The Stern Review*: N. Stern, *The Economics of Climate Change: the Stern Review* (Cambridge University Press, 2007), particularly 23–4. On the other hand, many economists deny that their discipline rests on foundations of ethics. Examples are the two reviews of *The Stern Review* published in the *Journal of Economic Literature*, 45 (2007), by W. Nordhaus (686–702) and M. Weitzman (703–24).

When the results of your acts are uncertain, on what basis should you decide what to do? You might at first think you should do what is most likely to have the best results. But that would be a bad mistake. You can see why by thinking about whether or not you should buy a fire-extinguisher. Your house is unlikely to catch fire. A fire-extinguisher costs money, and if you do not buy one but your house does not catch fire – as is most likely – that money will be saved. So if you base your decision on what is most likely to happen, you will not buy a fire-extinguisher.

However, that is not what you ought to do. If your house does catch fire and you have no fire-extinguisher, the result will be dire. This unlikely but possible consequence is so bad that (for most of us at least) it makes it worth buying a fire-extinguisher to prevent it. The lesson is that, when you are uncertain of the results of what you decide, your decision should not be based on what is most likely to happen.

This is an important lesson for climate change. When deciding what to do about climate change, the important thing may not be what is likely to happen. The important thing may be what is unlikely to happen, but may. The most likely result of increasing greenhouse gas is two or three degrees of further warming. But the unlikely possibility of eight or ten or twelve degrees might be a more important consideration, since its consequences would be catastrophic. I am not saying this is so; only that it may be so. We need at least to think about whether it is so. Is climate change a fire-extinguisher case?

The lesson so far is that we should not base our decision on what is most likely to happen. What is the correct criterion, then? It is expectation. More exactly, it is expectations of benefits and costs. This is the conclusion of our standard theory of right action in the face of uncertainty, which is known as 'expected value theory'. I shall not here try to demonstrate the truth of this theory; that requires some difficult theoretical work.[5] Nevertheless, I shall take its truth for granted.

The fire-extinguisher example does not demonstrate its truth, but it does illustrate how expected value theory delivers a correct result. Suppose you do not have a fire-extinguisher, so that if your house catches fire it will burn down. Your expectation of harm – given that you have no extinguisher – is the badness

---

[5] The main argument was first propounded by F. Ramsey, 'Truth and probability', in D. H. Mellor (ed.), *Foundations: Essays in Philosophy, Logic, Mathematics and Economics* (London: Routledge and Kegan Paul, 1978), pp. 58–100. It was developed by J. von Neumann and O. Morgenstern in *Theory of Games and Economic Behavior* (Princeton University Press, 1944), by L. Savage in *The Foundations of Statistics* (NY: John Wiley and Sons, 1954) and by R. Jeffrey in *The Logic of Decision*, 2nd edition (University of Chicago Press, 1983). These authors demonstrate that you should maximise the expectation of something generally called 'utility'. It takes a further argument to show that utility is a measure of value, so that you should maximise the expectation of value. This argument appears in J. Broome, *Weighing Goods* (Oxford: Blackwell, 1991).

of the house burning down multiplied by the small probability of its doing so. The badness is so great that, even multiplied by the small probability, it outweighs the cost of a fire-extinguisher. So according to expected value theory you should buy a fire-extinguisher. That is the correct conclusion.

Killing

We need to judge the value of the benefits and costs of different responses to climate change, and compare them together. This valuation must ultimately be based on ethical considerations. I shall next describe some of the ethical issues climate change raises. There are many. Some are easy to deal with and some hard. I shall mention some of the harder ones. These are issues of life and death.

Climate change will kill many people. It has various means of killing. The most obvious is the direct effect of heat. Because of climate change, heat waves are becoming more frequent, and heat waves are killers. A decrease in the number of cold waves is also expected, but cold waves do not kill people in such numbers as heat waves do.[6] There are also other lethal direct effects of weather, such as floods and storms. But direct effects kill comparatively few people in comparison with the effects of climate change that work less directly through disease and malnutrition. Climate change makes people more vulnerable to tropical diseases including malaria and diarrhoea. It also disrupts farming, which increases poverty and malnutrition, through either famines or chronic lack of food.

We need some idea of numbers. These are hard to come by, and estimates can only be extremely rough. It is not even easy to know how many deaths are already being caused by climate change, and predicting the number of future killings is harder. A few figures have been published. The World Health Organization (WHO) has estimated that in 2004 about 141,000 deaths were caused by climate change.[7] The Global Humanitarian Forum extended the WHO's methods more recently and reached the figure of 300,000 deaths caused by climate change in 2010.[8] They say this is a conservative figure. At the moment we have had less than one degree of global warming, and we are almost sure to get twice that much in the next few decades. I doubt we would be overestimating if we assumed that, from a few decades from now onwards, climate change will kill around half a million people a year. There is no clear

---

[6] IPCC, *Managing the Risks of Extreme Events and Disasters to Advance Climate Change Adaptation: Summary for Policymakers*, A Special Report of Working Groups I and II of the IPCC, C. B. Field *et al.* (eds.), (Cambridge, UK and New York, USA: Cambridge University Press, 2012), pp. 1–19, p. 11.
[7] WHO, *Global Health Risks: Mortality and Burden of Disease Attributable to Selected Major Risks* (Geneva: WHO Press, 2009), p. 50.
[8] Global Humanitarian Forum, *The Anatomy of a Silent Crisis: Climate Change Human Impact Report* (2009), www.ghf-ge.org/human-impact-report.pdf, last accessed 10 October 2014, p. 1.

end to all this killing. If it goes on for decades, it amounts to tens of millions of deaths attributable to climate change.

This is an important part of the harm that climate change will do. One of the benefits of controlling climate change will be to reduce the scale of this harm. In order to decide what we should do about climate change, we need to evaluate the benefits of acting and compare them with the costs. The costs are the efforts we have to put in: the sacrifices we bear to reduce emissions. These are things like travelling less, eating less meat, insulating buildings and diverting some of our consumption towards reducing emissions. The benefits include saving lives. How can we possibly compare that benefit against our more mundane sacrifices?

Economists are well used to making comparisons of this sort. For example, governments spend money on making roads safer. In doing so they are guided by just this sort of comparison. It is a routine of transport economists. They have a value for human life, measured in terms of money, and they compare the value of the lives saved with the cost of saving them. Health services make similar comparisons. For example the UK's National Institute for Clinical Excellence values a person's life at between £20,000 and £30,000 per year.

There is a lot to criticise in the way economists arrive at these numbers in practice, but I am not going to discuss their methods. Instead I shall discuss the principle – the whole idea of setting a value on life in terms of money. Many people find this idea horrifying. Life is sacred, they think, and we should not assign a money value to it.

What could we do instead? One possibility would be to treat lives as infinitely valuable. But that would be wrong. It would clearly be wrong to preserve lives at all costs. If a health service did this, it would not use any of its resources to improve lives in any other way than by extending them. It would do no hip replacements, for example, because hip replacements do not extend lives; they only save people from great pain. But clearly it is sometimes worth improving people's lives in this way.

Our lives are not infinitely valuable. You might nevertheless think it wrong to assign them a finite value in terms of money. Your view might be that not all good things are commensurable in value; their values cannot all be measured on the same scale. Climate change will destroy some of the beauty of nature, and this is a great loss. It will destroy many species of plants and animals. It will destroy the cultures of Arctic peoples. It would be a misunderstanding of these losses to assign them a monetary value that puts them on the same scale as not having foreign holidays and eating less meat. These goods are not all commensurable with each other, you might think.

There are indeed genuine incommensurabilities of value. Intuitively, we encounter them within our lives; we face them in many of the decisions we make for ourselves. Take a teenager who is choosing between a career in medicine and one in music. These two ways of life realise values that are very different from each other. It seems implausible that their values can be precisely weighed against each other. You might think that human life is similarly incommensurable in value with the mundane goods whose value can be measured in money.

But that would be a mistake. What is this value of life we are considering? It is the value of having a longer life rather than a shorter life. If your life is saved, you are not given some completely new thing; you are given more of what you have already. And what is the good of that? Well, living is good because good things come to you while you live. You get to have fun with people, visit beautiful places, paint pictures or look at them, go on holidays, eat nice food and all the other worthwhile things you do. These are just the mundane good things of life. The harm death does is to take these goods away from you; missing a foreign holiday is the same loss to you whether it happens because you are saving carbon dioxide emissions or because you die before you travel. Conversely, the good of life saving is to give people more of these mundane goods. Since the good of life saving is made up of mundane goods, it cannot be incommensurable with them. True, there may be incommensurabilities among the mundane goods themselves, but the value of life is not separately incommensurable with them.

The upshot of this is that, when we come to assessing what we should do about climate change, we shall not encounter ultimately intractable ethical questions surrounding the saving of people's lives. To be sure, there are difficult questions, but in principle they are ones we can handle with the methods of ethics we have.

Ethics of population

The real difficulty arises at the other end of life. There are two ways to add more life to the world. One is to extend the lives of already living people; the other to create more people. I now turn to the second way. Creating more people adds to the Earth's population, so I am turning to the ethics of population. Here there really are difficult problems.

They cannot justifiably be ignored when we think about how to respond to climate change. The growth of the Earth's population is one of the major causes of climate change. Reciprocally, climate change will influence the Earth's population. It follows that any action we take to limit climate change will also influence the population. This is the elephant in the room of climate change ethics. It is a huge ethical problem for climate change, but it is generally ignored.

How is population treated by people who think about climate change? Take William Nordhaus as an example. Nordhaus is an economist who has been working on climate change as long as anybody; he is one of the most important figures in the field. He created an 'integrated assessment model' called DICE, which he uses for comparing alternative approaches we might take to managing climate change. He uses this model in order to evaluate each of the different programmes we might follow.

Each programme involves putting a particular amount of effort into reducing greenhouse gas emissions, and distributes this effort over time in a particular way. For example, one programme puts a lot of effort into reducing emissions very soon. Another gradually increased the amount of effort over time. Nordhaus uses the DICE model to predict the effects of each programme and to set a value on each. His aim is to identify the best. He aims to 'optimise', that is to say.

This is a project within ethics. When Nordhaus sets a value on different programmes, he engages in ethical judgement. It requires the mathematical methods of economics to apply ethics to the extraordinarily complex problem of climate change, but these methods must be built on ethical foundations. Nordhaus himself is reluctant to recognise the ethical basis of what he does.[9] But in fact the 'objective function' he uses to evaluate alternative programmes within the DICE model embodies Nordhaus's own ethical theory. It reveals his ethics of population.

The function is:[10]

$$W = \sum_{t=1}^{T\,max} u[c(t), L(t)]R(t)$$

where

$$R(t) = (1+\rho)^{-1}$$

and

$$u[c(t), L(t)] = L(t)[c(t)^{1-\alpha}/(1-\alpha)]$$

The notation is as follows. W is the objective that is to be maximised – overall value, we may say. t is time, which is divided into discrete moments that are indexed from one up to $Tmax$. $c(t)$ is consumption per capita at time t. $L(t)$ is population at time t. $U[c(t), L(t)]$ represents the overall value of the world at time t. $R(t)$ is a discount factor that gives less weight to value at later times than

---

[9] W. Nordhaus, 'A review of the *Stern Review* on the economics of climate change', *Journal of Economic Literature*, 45 (2007), 686–702. See especially pp. 691–2.
[10] W. Nordhaus, *A Question of Balance: Weighing Options on Global Warming Policies* (Yale University Press, 2008), p. 205.

to value at earlier times. ρ is the discount rate. α is a parameter known as the 'elasticity of marginal utility of consumption'.

The formula

$$c(t)^{1-\alpha}/(1-\alpha)$$

stands for per capita well-being (Nordhaus would call it 'utility') at time t. Per capita well-being at a time is assumed to be a function of per capita consumption at that time, and the parameter α fixes the precise form of the function. The form does not matter here. The function determines *temporal* well-being: how well a person's life goes at the time. Temporal well-being must be distinguished from *lifetime* well-being, which is how well a person's life goes as a whole.

In Nordhaus's objective function, per capita temporal well-being at a time is multiplied by the population at that time. The resulting product is the total of temporal well-being at the time. This total is then multiplied by the discount factor, and added up across time to give the overall value that constitutes Nordhaus's objective. The resulting formula represents the ethical theory that Nordhaus assumes.

It shows that, in the ethics of population, Nordhaus adheres to a discounted version of 'total utilitarianism'. Total utilitarianism is the theory that the value of a world is the total of the lifetime well-beings of the people who live. Had Nordhaus set his discount rate at zero, the theory implied by his objective function would have been exactly total utilitarianism, with the added assumption that a person's lifetime well-being is simply the total of the temporal well-being she enjoys in her life. Actually Nordhaus assumes a positive discount rate, which is a complication. It implies that a person's lifetime well-being is a discounted total of her temporal well-beings – discounted in a way that gives less weight to later times in her life than to earlier times. It also implies that the lifetime well-beings of people who live later in time are less valuable than the lifetime well-beings of those who live earlier. Nevertheless, it is fair to treat Nordhaus's implicit theory as a version of total utilitarianism.

Nordhaus did well to multiply per capita temporal well-being at each time by the population. Had he not done so, his implied ethical theory would have been that the value of the world is the discounted total of per capita temporal well-being at each time. This theory is absurd. It implies it would be better if everyone whose temporal well-being is below the average were to die.

What difference does Nordhaus's ethical theory make to his conclusions about climate change? Since Nordhaus multiplies well-being per capita by the number of people, the effect is to give more weight in his valuations to the quality of life at times where there are more people in the world. That is to say, it gives more weight to the future. This makes a significant difference to his conclusions. Most of the action in the economics of climate change revolves around balancing the interests of the future against the interests of the present.

Climate change policy, looked at from this point of view, is a matter of the current generation's sacrificing its own well-being for the sake of promoting the well-being of future generations. By adopting total utilitarianism, Nordhaus is giving some more weight to the future. As a separate matter, Nordhaus applies a discount factor to well-being, which is to give less weight to the future. His total utilitarianism counteracts his discounting to some extent.

This is a significant consequence. But the full importance of population ethics in climate change does not emerge in Nordhaus's work, because Nordhaus treats the growth of population as exogenous. He does not consider the effects of climate change, and of policies that respond to climate change, on the Earth's population. If he did allow for endogenous population, he would be inclined to favour policies that add people to the world. This is a natural consequence of total utilitarianism. Total utilitarianism implies that, for any given level of well-being (provided it is counted as a positive rather than negative level), the more people who enjoy well-being at that level the better the world is.

Most people find total utilitarianism, whether discounted or not, unattractive. Most of us care about how things go for the people there are. We want to make life better for them. We do not attach value to the number of people. We think that adding new people to the world is not in itself a good thing, even if those people will be well off. If well-being is added to the world by making a person who already exists better off, that is a good thing. But if well-being is added to the world by creating a new person to enjoy it, that is not in itself a good thing. Prolonging the life of an existing person is normally a good thing because it normally adds to the goodness of the person's life as a whole, but creating a new life is not a good thing, even though both these actions bring it about that more good life is lived. In the words of Jan Narveson, 'we are in favour of making people happy, but neutral about making happy people'.[11]

What was the Chinese government thinking of when it introduced its one-child policy, in an attempt to reduce the growth of the Chinese population? It was thinking of improving the standard of living of Chinese people. It believed it could promote their standard of living by holding back the population, so as to reduce the pressure on resources. I am sure it did not think it a consideration against its policy that it would reduce the total amount of well-being enjoyed by Chinese people, because there would be fewer people to enjoy it. In this respect, many of us think like the Chinese government.

When we see a couple who decide not to have a child, although they could have a happy child if they chose, we mostly do not think they are giving up an opportunity to make the world a better place. If they were, their decision would

---

[11] J. Narveson, 'Moral problems of population', *The Monist*, 57 (1973), 62–86.

Climate change: life and death   195

be morally dubious: normally we are morally required to improve the world if we can. But we mostly think there is nothing wrong with their decision.

In sum, I think most of us are gripped by the intuition that adding a person to the world is morally neutral. It makes the world neither better nor worse. I call this the 'intuition of neutrality'.

So not only does Nordhaus's value function embody an ethical theory, it embodies one that does not accord with common intuition. I am not saying that common intuition is right and Nordhaus wrong. Indeed, it will emerge that I am more on Nordhaus's side than against it. I am simply emphasising that the ethical theory of population implicit in Nordhaus's work is significant. It cannot be taken for granted.

Climate change, population and the chance of catastrophe

How does our population ethics – whether total utilitarianism or the intuition of neutrality or something else – affect the judgements we should make about climate change? It depends on how climate change will affect the world's population, and that is not yet clear for climate change at the level that is thought likely: a few degrees. I do not know of any predictions of the demographic effects of climate change. I do not even know whether likely climate change can be expected to increase or decrease the Earth's population. We do not know how to make population endogenous to our models, and we cannot tell how doing so would affect our conclusions about climate change.

But for severe climate change, which is far beyond the likely level, things are different. Severe climate change will be catastrophic. It will drown our coastlines under the sea; it will wipe out many of our fresh-water supplies; it will make it much harder to grow crops; it will cause conflicts over our diminished resources. It will make the world much less habitable for human beings. It will inevitably cause a collapse of our population. It may even cause our extinction. What should we think of that?

This is an important practical question. Severe climate change is unlikely. However, I explained in the section 'Expected value' that unlikely possibilities are not necessarily unimportant. Indeed, they may be the most important consideration for some decisions, as they are for buying a fire-extinguisher. It depends how bad the unlikely event is. If it is extremely bad, its badness may outweigh the small chance of its occurrence. We need to know whether the chance of severe climate change is like the chance of your house burning down. We therefore need to think about how bad its results would be.

It would be very bad in one obvious way. If severe climate change causes our population to collapse, it will not do so in a nice way. The collapse will happen through horrific events: starvation, drought, disease and war. Huge numbers of people will suffer, and many will die before their time. These are terribly bad consequences that will result from severe climate change. Their badness is independent of population ethics. Suffering and killing is bad on any theory.

However, surprisingly perhaps, all this dreadful badness is not enough to make severe climate change like the case of the fire-extinguisher. The numbers show why. Suppose, say, there is a one-in-a-hundred chance that severe climate change will kill billions. I do not think we should put the chance at more than this. Then the expected number of deaths is billions divided by one hundred, which comes to tens of millions. But I explained earlier that tens of millions of people are going to be killed even by the degree of climate change that is likely. So this small chance of killing billions does not dominate the calculations in the way that the chance of a fire dominates the calculations about a fire-extinguisher. It is a dreadful harm, but because of its small probability its expectation of badness is of the same order as the badness of killing we must expect from likely climate change. To dominate, it would have to be much greater.

Besides killing all those people, severe climate change will have another effect. It will prevent the existence of other people. Since the population will collapse, many people who would otherwise have existed will actually not exist. If we become extinct there will be no more people again, ever. Whether or not there is complete extinction, we may say that there will be many 'absences'.

The number of absences severe climate change will cause is likely to be much larger than the number of people it will kill. To take the most extreme case, suppose humanity becomes extinct. Had this not happened, humanity might plausibly otherwise have survived with a population of billions for a hundred thousand more years, renewing its population every hundred years or so. This puts the number of absences in the trillions. If absences are bad, the sheer number of them might well be a harm that makes severe climate change like the case of a fire-extinguisher.

That depends on how bad the absence of a person is, which is a question for the ethics of population. What you think about it will depend on your theory about the value of existence. I said that most of us intuitively do not value the existence of people for its own sake. We should therefore not disvalue the absence of people. Those of us who are gripped by this intuition of neutrality should not think that a collapse of population or even the extinction of people is in itself a bad thing. Many of us are not pleased by the fact that population is growing; we should therefore not be displeased if it shrinks. So from our point of view, the catastrophe caused by extreme climate change will be the huge amount of killing it will cause, not the subsequent absence of people. I have explained that the killing is not enough to make climate change like the case of the fire-extinguisher.

I think many people will find this thought a challenge to the intuition of neutrality. When we are gripped by the intuition of neutrality, we are not normally thinking about such extreme consequences of it. Thinking about a collapse of our population, or our extinction, may incline us to change our minds. Many of us also have a conflicting intuition that the extinction of humanity would be a dreadfully bad thing. This should make us think again about the intuition of neutrality.

This does not mean we have to reject it. There are separate considerations that could explain why the extinction of humanity would be a dreadfully bad thing. It may be that humanity has a value over and above the value of the people who make it up. We often think this about other species of life, and humanity has qualities that plausibly make it a particularly valuable species. We have great achievements to our credit: language, science, art, rationality, morality, civilisation. These may have a value of their own, which makes them worth preserving and developing.

To take this view, we do not necessarily have to think hubristically that they have a value from the point of view of the universe. We might think alternatively that they have a value from the point of view of humanity, viewed as a species or a collective. In the same way, a person's continuing life has a value for the person herself. Her continuing life is a good thing for her: better for her than dying. Similarly, continuing the human species may be a good thing from the point of view of humanity itself, better for us than extinction.

Rejecting the intuition of neutrality

So we are not forced to give up the intuition of neutrality by the contrary intuition that the extinction of humanity would be a bad thing. However, we should give up this intuition anyway. I do not say this on grounds of any contrary intuition, but on grounds of argument. I cannot present the full argument here, but to illustrate the way it goes I shall describe the beginning of it.[12]

The intuition of neutrality is the intuition that adding to the world's population makes the world neither better nor worse. To take a particularly simple case, compare two possible worlds. One is the actual world. The other is exactly the same, except that there is an extra person added to the population. Everyone else in the other world is exactly as well off as she actually is. In practice, some people would undoubtedly be affected by the existence of the extra person – her parents for instance. But imagine that anyone who is affected for good or ill by

[12] See Broome, *Weighing Lives*, chs. 11 and 12.

the extra person's existence has her well-being improved or damaged in some exactly balancing way, so that the net effect on her well-being is nothing. The intuition of neutrality implies that the world where an extra person exists is neither better nor worse than the actual world where she does not exist.

The intuition splits at this point into two versions. According to the strong intuition of neutrality, as I shall call it, the two worlds in this example are equally good. According to the weak intuition on neutrality, they may be incommensurate in value. I shall show that the strong version is false.

The strong version claims that the two worlds I have described are equally good. Now take a third world. It is exactly like the second one, containing the same people including the same extra person. But in this third world, the extra person is better off than she is in the second world. Everyone else is exactly as well off as she is in both the first world and the second. Think about the relative goodness of the second and third worlds. These worlds contain exactly the same people, and all but one of the people they contain is equally well off in each. That one person is better off in the third world. So the third world is obviously better overall than the second: it is better for one person and equally good for all the rest.

Now think about the relative goodness of the first and the third world. The only difference between these worlds is that the third contains a person who does not exist in the first. Everyone else's well-being is the same in the two worlds. The strong intuition of neutrality implies these two worlds are therefore equally good.

We know already that the strong intuition implies that the first and second worlds are equally good. So the second world and the third world are both equally as good as the first, according to the intuition. This means they are equally as good as each other. But that is not so; we know already that the third is better than the second. So the strong intuition of neutrality implies something false. It is therefore false itself.

That is a conclusive argument against the strong intuition of neutrality. My argument against the weak intuition is more complicated, and I shall not present it here.[13] I have to admit that it is less conclusive than the one I have given against the strong intuition.

We mostly have the intuition of neutrality. But thinking about the possibility of extinction gave us one reason to doubt it. And there is also the good argument against it that I have just described. We should give up this intuition.

---

[13] See Broome, *Weighing Lives*, ch. 12.

What to do?

What alternative account of the value of population do we have? I have mentioned total utilitarianism, which values the total of all people's well-being. Total utilitarianism implies that a collapse of population would be an enormously bad disaster. It would cause the loss of all the well-being of all those trillions of future people who would be absent. But most people find total utilitarianism implausible too. Some other ethical views about the value of population are also in circulation. But all of them are subject to strong objections of one sort or another.[14] At present the state of population ethics is confused. It is much debated in the philosophical literature, but nothing close to a consensus has emerged.

This presents a serious problem for policy making about climate change. Severe climate change is a real possibility. If it would be truly catastrophic, this possibility will make climate change like the case of a fire-extinguisher. Though unlikely, it would be so bad that it should dominate our planning. This is the view that the economist Martin Weitzman has been propagating for some years.[15] But if it should really dominate our planning, that will only be because it causes a collapse of population, or even extinction. So we cannot judge the badness of a catastrophe without judging how bad a collapse of population would be. Yet we have no secure basis for doing so.

We mostly have the intuition of neutrality. I suspect this explains why people have thought so little about the effects on population when they assess climate change policies. But the intuition of neutrality is false. And at present we have no secure ethics of population to put in its place.

What can be done about this problem for policy making? At a theoretical level the problem is particularly intractable. All the theories we can come up with about the value of population are subject to strong objections. But at a practical level it is not so far different from many problems of policy. Because population ethics is so intractable, philosophers disagree deeply about it, but disagreement is ordinary. We are used to disagreements over many issues, among experts and among the public. Yet decisions have to be made, and they get made despite the disagreements.

A society cannot hope to arrive at agreement about everything before it acts. We therefore have a way of making decisions that does not require agreement. One purpose of a democratic political system is to mediate between people's different opinions. However the moral philosophy of climate change progresses, it will actually be the political system that determines what we shall

---

[14] G. Arrhenius, *Population Ethics* (Oxford University Press, forthcoming) explains the profound difficulties that all theories face.

[15] M. Weitzman, 'On modeling and interpreting the economics of catastrophic climate change', *Review of Economics and Statistics*, 91/1 (2009), 1–19.

do about climate change. Provided the system works well, we should welcome that fact.

To work well, a democratic system requires people to be well-informed participants. This gives a role to moral philosophy in the democratic process. Philosophers need to participate in the public debate and provide guidance to people in making up their minds. When, as philosophers, we think we know the answer to a question, our job is to put our answer into the public domain, and defend it with arguments as well as we can. When we are unsure of an answer, we should explain what we see as the alternatives, and what merits they have. When we disagree about an answer, we should argue the question out. We can at least offer the public ways of thinking about the questions.

As a contribution to the debate about climate change, one thing I have been trying to do is point to the elephant in the room. In making our judgements, we need to recognise the importance of population, and think about its value.

11   What we have done ≠ what they can do

Benjamin Hale

> We knew the world would not be the same. A few people laughed, a few people cried. Most people were silent. I remembered the line from the Hindu scripture, the Bhagavad-Gita. Vishnu is trying to persuade the Prince that he should do his duty, and to impress him, takes on his multi-armed form and says, 'Now I am become Death, the destroyer of worlds.' I suppose we all thought that, one way or another.
> Robert J. Oppenheimer

> My god. What have we done?
> Robert Lewis

When Robert Lewis, co-pilot of the Enola Gay, jotted the above famous words in his flight notebook, he encapsulated a sentiment much more general than the otherwise specific earth-shattering instance to which he was referring. What he asked was simple and plain. Nothing as poetic or trenchant as Robert Oppenheimer's chilling reflections. And yet, it is nevertheless profound and resonant: What have we done?

Each of these four words invites further consideration: not just about what event we have caused, but also about what we, humans, have done to cause that event. In the above instance, Lewis is referring to the bombing of Hiroshima, which famously inaugurated the nuclear age. But the question could easily be directed at a host of anthropogenic issues, including, of course, climate change.

The simple fact is that climate change will dramatically affect the planet, impacting not only precipitation, drought and sea ice, but also the distribution of pests and disease, the success of agriculture, the shape of urban development and the survival and vitality of plant and animal life. As is no doubt clear from the preceding chapters in this volume, much of the climate justice discussion centres on how climate change will deleteriously impact humans. But climatic shifts will perhaps be that much more pronounced for those that are ensconced – in a way, trapped – within unique ecological niches. As the climate destabilises, shifts and establishes new equilibrium states, so-called 'novel ecosystems' will begin to emerge.[1] Regional plant and animal life will be forced to adapt or die,

---

[1] E. C. Ellis, 'Anthropogenic transformation of the terrestrial biosphere', *Philosophical Transactions of the Royal Society A: Mathematical, Physical and Engineering Sciences*, 369

accelerating extinction rates and, arguably, reducing biodiversity and homogenising populations. Such an outcome is worrisome, and not simply because there may be spill-over impacts on humans. It is a post-industrial tragedy the scale of which is only mildly comprehensible when put in the context of the nuclear age.

Naturally, one of the easiest entry points to any robust and demanding environmental ethics is through the demonstration of the moral status of non-human nature. Through the decades, many have attempted such a demonstration, all with varying degrees of success. I cannot cover these views here. Any animal or environmental ethicist will tell you, however, that one of most pernicious objections to the view that non-human animals, or more broadly, non-human nature, deserves some moral consideration is that people are rational and animals are not. Partly as a consequence of this, animal and environmental ethicists have offered substantial refutation of this point. We do not deny human children rights, for instance, so the mere fact that an entity is not rational should not be enough to discount it entirely. Nevertheless, the response persists, and remains an imposing obstacle to those who wish to argue on behalf of non-human nature.

In this chapter, I shall approach the same question somewhat unconventionally. Rather than approaching the question of obligations through the lens of moral status – regarding the scope of morality – I will instead approach the question from the standpoint of justice. More specifically, I aim to raise the question of whether, when raising justice questions, we can make sense of obligations not only to human nature, but also to non-human nature. I shall suppose that we do have such obligations; and I shall suggest that we have these obligations because we are responsible for the messes that we make. That is, I believe I can address the climate justice question by turning attention to the anthropogenicity of climate change. My fundamental claim is that our obligations to address the problem of climate change are rooted in the unjustness – by the unjustified-ness – of our actions.

Moral considerability

For at least the past forty years environmental ethicists have sought to wedge their way into moral theory by appeal to values, principles, standards and assumptions common throughout the ethical encyclopedia. They have argued using all variety of strategy, typically by asserting that nature is valuable for some property inherent to it, that humans are a part of nature, that commonly

(2011), 1010–35; E. Higgs, 'Changing nature: novel ecosystems, intervention, and knowing when to step back sustainability science', in M. P. Weinstein and R. E. Turner (eds.), *Sustainability Science: the Emerging Paradigm and the Urban Environment* (New York: Springer, 2012); R. J. Hobbs, E. Higgs and J. A. Harris, 'Novel ecosystems: implications for conservation and restoration', *Trends in Ecology and Evolution*, 24 (2009), 599–605.

assumed lines demarcating man from nature are illegitimate, that stewardship of nature is a virtue, and on and on. These environmental ethicists have rummaged through the history of philosophy in search of some method to argue persuasively that animals, plants and other natural objects are worthy of our moral consideration – that they are, in other words, morally considerable.

In earlier work, I argued for a more deontological approach to moral considerability.[2] I argued that moral considerability should be understood as a question not so much about status, but rather about obligations: a question that inquires into the presuppositions of practical reason in order to determine what obligations rational agents have to non-human others. I proposed that, rather than identifying qualifiers for moral patienthood, moral considerability makes better sense when understood as narrowly trained on whether and why an agent has responsibility to assess a set of considerations regarding entities in the world. Approaching the problem of moral considerability in this deontological way, I claimed, obviates the problem of identifying specific attributes in patients by limiting the question of moral considerability to agents.

My early approach was to unpack the distinctions implicit in the idea of moral status, which lurks under the surface of the moral considerability query. I suggested that moral status intercalates at least three related questions into one. First, it raises a question of what or which entities must we *consider*, and why must we consider them. (This, I argued, more appropriately captures the idea of moral considerability.) Second, it raises a question that follows directly from considerability: if we must consider some thing, then what must we consider *about* that thing? Third, it then invites us to ask: given the relevant considerations, how *much* must we consider each consideration or, differently put, how much *weight* must we give to these considerations? The question of moral status as traditionally conceived attempts to answer all three questions at once, telling us which entities we must consider by virtue of which attributes and according to the degrees to which those attributes are relevant.

We are better off, I believe, if we keep these status questions distinct, and instead limit our inquiry to the narrower questions of moral considerability, moral relevance and moral significance.

The way I read it, environmentalists face an exclusively deontological problem when they ask themselves about the status of environmental entities. How we treat the environment is *our* problem, and not straightforwardly the

---

[2] B. Hale, 'Technology, the environment, and the moral considerability of artifacts', in J. K. B Olsen *et al.* (eds.), *New Waves in Philosophy of Technology* (London: Ashgate, 2008); B. Hale, 'Moral considerability: deontological, not metaphysical', *Ethics and the Environment*, 16/2 (2011), 37–62.

problem of the animals, trees or mountains. If instead we conceive of moral considerability not as a question about the *rights* of an animal, group of animals, or natural entity, but rather as a question about how one should go about *considering* the animal or entity, then we can avoid some of the apparent conflicts between the many ethical positions.

The pivotal assumption here is that the trappings of agency come with a heavy price tag: that being moral agents, being morally responsible, means adhering to certain rules of behaviour – moral rules. These rules are derivable first from our communicative interactions with other rational agents, and then from our own reflections on the principles that underwrite these interactions. The burden is on us – human animals with voices and minds, in other words – to approximate the morally binding rules and principles that are already in play in human–non-human relations.

In other papers, I and my colleagues also argued for obligations to non-human nature by starting from an intuition that many of us hold: that we must right prior wrongs.[3] We suggested that many of the more traditional approaches to arguing for ecological intervention, complicated by the unpredictable vicissitudes of climate variability, fall subject to considerations that threaten to derail intervention altogether.

In many of these arguments, we assumed, but did not defend, the argument from reparation, which in the below formulation has a climatological emphasis:

Argument from reparation: because climate change, and the consequences stemming therefrom, is a predicament of our own making, we have an obligation to assist nature with adaptation.

What we claimed, essentially, was that all we really need to get intervention ecology off the ground is a simple acknowledgment of our role in bringing about climate change.

Typically, one might encounter such an argument and assume that an obligation of reparation must require a victim or a patient that can be made whole again. So if Jones wrongs Smith by harming him, Jones must repair the damage that he has inflicted upon Smith, presumably in order to 'make Smith whole again'. But I want to suggest that obligations of reparation do not require such a victim, that we can have an obligation to undo what we have done simply

---

[3] B. Hale. 'Getting the bad out: remediation technologies and respect for others', in J. K Cambell *et al.* (eds.), *The Environment: Topics in Contemporary Philosophy*, Vol. 9 (Cambridge, MA: MIT Press, 2012); B. Hale, 'Polluting and unpolluting', in M. Boylan (ed.), *Environmental Ethics*, 2nd edition (Hoboken, NJ: Wiley-Blackwell, 2013); B. Hale, 'Can we remediate wrongs?', in A. Hiller *et al.* (eds.), *Consequentialism and Environmental Ethics* (New York, NY: Routledge, 2013); B. Hale, A. Hermans and A. Lee, 'Adaptation, reparation, and the baseline problem', in M. Boykoff and S. Moser (eds.), *Toward Successful Adaptation: Linking Science and Practice in Managing Climate Change Impacts* (London and NY: Routledge, 2013), pp. 67–80; B. Hale, A. Lee and A. Hermans, 'Clowning around with conservation: adaptation, reparation, and the new substitution problem', *Environmental Values*, 23 (2014) 181–98.

because we have done it. That's a fairly strong thesis, and I suspect upends much conventional thinking on the nature of wrongdoing. Nevertheless, I think it's true, and I think the way to gain access to that thesis is by rethinking our approach to justice. Importantly, I mean 'justice' here less as embodying a certain set of obligations and more as a procedural obligation of justification, as recently advanced by theorists Jürgen Habermas and Rainer Forst.[4]

What I have been driving at in this work is that the troubling problem for environmental ethics isn't so much whether and what sorts of entities ought to be included in our moral figurations, but rather whether the actions that we take can be justified. I have suggested that the wrongness of our actions with regard to non-human nature rests not in harming morally considerable or intrinsically valuable entities, but rather in taking actions that are wanton, reckless, inconsiderate, selfish, or even just blind and naive. What I mean by this is that many of these actions – the actions that we take that, cumulatively, cause climate change – are not subjected to the scrutiny of other affected parties, and in virtue of this, are not put through a justificatory process. A fundamental piece of this puzzle therefore requires demonstrating that individual or collective actions – throwing a ball, gassing up the car, shooting a pigeon, waging a war, geoengineering the planet – are distinct from natural events in such a way that invites the exchange of reasons. It further involves demonstrating that an imperative to justify our actions is wide enough to accommodate claims about the wrongness of abusing animals, encroaching on habitat, or destroying the planet.

I do not have the space in a chapter of this length to cover the full range of metaethical or even normative contours of such a position. Far more established philosophers than I are doing this work. Building on the work of Jürgen Habermas, for instance, the German theorist Rainer Forst has done some exciting work in recent years to elaborate a related position, though one more directly associated with a right to justification and not with environmental questions per se.[5] Nevertheless, I think his view offers a fruitful pathway for the environmental ethicist.

I would like here only to give a brief outline of the intuitions that motivate Habermas and Forst in order to loop their discussion into the climate justice discourse. Below I aim to defend the view that environmental problems aren't fundamentally about establishing the moral standing of nature – whether

---

[4] J. Habermas, *Truth and Justification* (Cambridge, MA: MIT Press, 2003); J. Habermas, *Between Facts and Norms* (Cambridge, MA: MIT Press, 1994); J. Habermas, 'Discourse ethics', in C. Lenhardt and S. W Nicholson (eds.), *Moral Consciousness and Communicative Action* (Cambridge, MA: MIT Press, 1991); R. Forst, *The Right to Justification: Elements of a Constructivist Theory of Justice*, J. Flynn (trans.), (New York: Columbia University Press, 2012); R. Forst, 'The justification of human rights and the basic right to justification: a reflexive approach', *Ethics*, 120 (2010) 711–40.

[5] Forst, *The Right to Justification;* Forst, 'The justification of human rights and the basic right to justification'.

individual non-human animals, or aggregations, communities, or systems of animals – but instead altogether a question of whether the actions that we have taken – what we have done – are the kinds of actions that we can justify.

Approbation and disapprobation

P. F. Strawson's groundbreaking piece 'Freedom and resentment' is ostensibly and fundamentally a paper about free will, and mostly read as a response to sceptics about free will.[6] It also offers, however, a stepping off point for interrogating our responsibilities for environmental harm and devastation. Among other things, Strawson assesses the kinds of interpersonal emotions that follow in the wake of bad actions. He explains that our sense of indignation at guilty parties is an essential component of our psychology, of the social fabric, of understanding one another. It is his view that we plainly recognise a distinction between human-caused and naturally caused action, and grow indignant when we observe that some bad action is human-caused.

Strawson takes it as an intuitive, 'commonplace' fact that if another human intentionally does something terrible to us, we experience resentment towards that person. If, say, someone intentionally trips me, I might be upset by this. I grow upset because the tripper has acted intentionally, by employing his will, and he has wronged me. Compare this with our feelings about human-caused actions that were undertaken unintentionally. If instead the tripper immediately explains that he had not intentioned the accident, or that he has no control of his leg, or that he was trying to avoid being bitten by a snake, I might be less inclined to blame him, less inclined to feel disapprobation towards him.

These 'reactive attitudes', as Strawson calls them, are a fundamental part of who we are, of how we respond to human-caused actions. For Strawson, they are fundamental to our psychology. When evaluating an action, he explains, a person can take either the 'objective attitude', and think of the action as an event, or the 'reactive attitude', and think of the action as emanating from a free will.

What Strawson encourages us to ask is whether we would be any better off if we were to become hard-core sceptics about freedom of the will; whether, in fact, we could even actually become sceptics about freedom. His answer is that we would be much worse off; and that we would not be able to purge ourselves of the reactive stance if we tried. To give up the reactive stance would mean giving up our humanity, giving up practical reason. There is a consequence to accepting the thesis of determinism, he claims: giving up the reactive attitude will affect how we reason practically, not theoretically.

[6] P. Strawson, 'Freedom and resentment', *Proceedings of the British Academy*, 48 (1960).

The important point here is that, for Strawson, these reactive attitudes are essential facts of our psychology. What they reveal to us are our normative expectations; our expectations about what should and should not be; about what is justified.

Jürgen Habermas explains more thoroughly the nuances of Strawson's 'phenomenology of the moral'. He suggests that Strawson's contribution is important for the following three reasons: it demonstrates (1) that 'moral phenomena are grasped only in the performative attitude [which is Habermas's pragmatic equivalent of Strawson's reactive attitude] of participants in interaction'; (2) that resentment and 'personal emotional responses point to suprapersonal standards for judging norms and commands'; and (3) that 'moral-practical justification of a mode of action aims at an aspect different from a feeling-neutral assessment of means–ends relations, even when such assessment is made from the point of view of the general welfare'.[7]

What Habermas points out about Strawson is couched in a lingo that may seem impenetrable at first. However, the points are the following: (1) When it comes to evaluating actions, moral phenomena – that is to say, our experiences and use of terms such as 'worth', 'value', 'good' and 'bad' – can only be understood from the perspective of those who are caught up in social interactions. It simply does not make sense to speak of worth and value apart from our social interactions, because we are the only beings that assess worth and value claims. (2) The fact that we get irate about the way in which things go when people do nasty stuff to us or to others points to a standard of evaluation that stands apart from our personal emotional response. If I get irate at you for beaning me with a baseball, when I am the batter and you are the pitcher, I get irate at you because you have violated some standard of fair play, some normative expectation. I am not just irate because you've done something that I don't like or want. And finally, (3) when we seek to justify particular claims to actions, our appeals extend to these supra-personal standards, to standards that are set apart from simple means–ends relations. You as pitcher cannot simply write off your actions as somehow acceptable because you 'wanted to take me out of the game'. There are acceptable rule-governed methods for taking me out of the game, as when an outfielder catches a pop-fly; and then there are unacceptable methods for taking me out of the game, which might involve causing me physical injury, as when Tonya Harding clobbers Nancy Kerrigan's kneecaps just before the big showdown. Habermas essentially denaturalises Strawson's account. Rather than locating reactive attitudes as essential elements of our psychology – a point that is plausible, if contentious – he proposes that it is an essential feature of communicative pragmatics that we attribute autonomous justificatory potential to our

---

[7] Habermas, 'Discourse ethics'.

interlocutor. Like Strawson, he proposes that any time we are involved in communication, we can take either one of two attitudes – either the objectivating attitude or the performative attitude; and that this is an essential feature, it is constitutive, of language. If we take the objectivating attitude, we might be inclined to treat our interlocutor as an object – as value-free. Thinking of our interlocutor as an object might enable our strategic manipulation of them. If, on the other hand, we adopt the performative attitude when we are involved in communicative interaction, we must treat the speaker as more than an object – as a reason-giver, as a reflective endorser, as an end-in-itself.

Now consider a case:

If Abercrombe pushes Smith over the ledge of a building, and Smith falls to his death, we most certainly feel compelled to hold Abercrombie responsible for his action. We might even feel indignation at Abercrombe. We feel this way until we can make better sense of what it was that caused Abercrombe to act in such a thoughtless and malicious manner.

If instead Bartholomew pushes Jones near the ledge of a building, and the ledge collapses under Jones's weight, we feel somewhat less indignant, since we are less certain about holding Bartholomew responsible for the event. We might hold the building engineers or the building manager responsible for the collapse of the ledge, but we cannot hold Bartholomew responsible for the bad occurrence in the same way that we can hold Abercrombe responsible. It is true that the outcome was terrible only because Bartholomew initiated a chain of events that precipitated in the death of Jones. But the bad turn of events was probably not what Bartholomew had intended, nor presumably was it what he had wanted, and so our reaction is considerably calmed by this bit of knowledge.

Part of our differing reaction to Abercrombe and Bartholomew clearly rests, then, in our understanding of their intentions. That is, it appears clear that Abercrombe *willed*, and took action, to ensure that his victim fall over the edge. With Bartholomew, we are less certain about the role of his will. Perhaps Bartholomew was just joking around, as one might joke around with a friend on a subway platform. Strawson would want us to acknowledge that we have a natural, psychological response to Abercrombe and Bartholomew; and that we would lead deficient lives if we were to accept the thesis of determinism. But Habermas calls us back into the world of interaction, and specifically into the world of communicative interaction. He encourages us not to speculate into the messy black-box of human psychology, not to focus exclusively on intentions but rather just to look at the sorts of excuses and non-intentionalistic reasons that can be offered between speaker and hearer on behalf of or by the perpetrator.

Habermas therefore encourages us to recognise that these reactive attitudes have a communicative counterpart: a counterpart that is not rooted in a psychological supposition about human nature, but rather in the kinds of excuses that can be offered on either Abercrombe's or Bartholomew's behalf. Standard philosophical parlance here might require us to distinguish between motivating reasons and justificatory reasons, since motivating reasons are allegedly the reasons that an agent 'has' or 'holds' for taking the action (and thus takes the action on the basis of this reason); whereas the justificatory reason might be a reason that is either known or unknown to the agent but that nevertheless justifies the action. But Habermas's push away from Strawson's psychological account essentially turns away from motivating reasons to appeal to justificatory reasons.

Bartholomew may be able to get us to settle down in our disapprobation relatively easily, as there are many ways in which we can imagine exculpating him. Yet there are also some circumstances in which we might exculpate Abercrombe. If Abercrombe can provide us with an adequate justification or excuse for his actions, we might be equivalently calmed in our reaction to him. If Abercrombe offers these explanations, such as 'He was trying to kill me', or 'I was under mind-control by Bartholomew', or even if we learn through some third party that Abercrombe has a deep psychological disorder – we might in these cases forgive them.

In assessing the actions of rational beings in this manner, we are able to determine whether the individual was *justified* in acting, whether the individual's will is responsible for moving him to take the successful action that he took. Importantly, on this conception, it is communicative uptake that establishes whether Abercrombie was justified or not. That is, independent evaluators of Abercrombie's reasons determine, by assessing the strength of Abercrombie's claims, whether his position is justified. In this way, justification isn't a status so much as a process; and whether an act is justified depends not on some set of ideal conditions, but rather on whether the reasons have been intersubjectively validated. What is important about Strawson's observation is that the reactive attitude is an attitude that is *inextricable from* our experiences of other human beings, and this is so for Habermas as well.

But is it the case that these reactive attitudes are limited to our interactions with linguistically competent, and thus, rational, interlocutors, as Habermas suggests? To a certain extent, yes. For if an interlocutor can provide us with a justification of their actions, then we are liable to forgive them. If not, then we are liable to become frustrated, to blame them, to resent them, and to find them culpable, and so Habermas's denaturalisation of Strawson works perfectly well.

Freedom and nature

What of cases where there is no linguistic or symbolically mediated interaction? Is it not the case that even here we can provide and find justifications? And is it not further the case that it is a constitutive supposition of communicative freedom that agents be able to distinguish between naturally caused events and deliberate actions? Is it not the case that we must always ask ourselves whether such and such an event qualifies as a *deliberate* action? And then, and only then, whether the action was justified?

Return to Strawson's account for a moment and consider the following example:

Suppose we are standing again on the self-same roof beside Abercrombie and Bartholomew. If a lightning bolt shoots out of the sky and fries Cornelius on this roof, we may be quite startled by this. We may seek an explanation. We may inquire into the mechanism that precipitated such a curious natural event, for lightning strikes that zap nearby acquaintances are reasonably uncommon, and a mechanistic explanation would help put us at ease about the recurrence of such an event. We may say, 'Shucks, Cornelius, looks like you picked the wrong day to wear that aluminum foil blazer'. Or we may at the same time explain the action in terms of intentionalistic metaphors. We may say, 'That lightning bolt *saw* in Cornelius a conduit to the Earth'. We may even frame the event in personified terms: 'Somebody up there is furious with Cornelius!' This is the mythological-religious worldview; the one that we, post-enlightenment deterministocrats, have pretty much demystified.

Most of the time, we don't ever get this far. We understand fairly well the mechanism that we call lightning, and so we also understand that the intentional stance in this case applies only metaphorically. But it is our *understanding* of the causes that stimulates our reaction to the event. This procedure extends on down to some of the most dramatic and heart-wrenching of natural events.

If a tsunami destroys villages along the coast of Thailand, killing 200,000 people, we may have reason to lament such a tragedy, to be concerned about the victims and to shake our heads in horror at the event. It is a calamity, plain and simple. We may even classify the event as 'bad'. We may explain it, however, in terms of tectonic shifts, or escaping tides, or even Nero's anger. In doing so, we explain the event by way of either intentionalistic or mechanistic description. Mechanistic descriptions are relatively common, and perhaps even 'true', for events like tsunamis; and so we can quickly avoid getting our moral hackles up about the event, even though we accept the event as 'bad'.

If, on the other hand, a weapon of mass destruction – a nuclear weapon, say – detonates off the coast of Thailand, killing 200,000 people, we have reason to be considerably more disturbed by the event, to be morally infuriated. We have reason to be disturbed by the event because we have a sense that the weapon did not detonate itself, but that there was some responsible agent behind the detonation. We might classify this event as 'bad' as well. But what is important is that we are forced into the intentionalistic description because there is no suitable naturalistic or mechanistic description.

What's the difference between these positions? How can we ascertain that there is a difference? After all, what we're really looking at are only two equivalently devastating events, both of which would not have been possible without nature's immense storehouses of energy.

One first thought is that the distinction between the two events rests in the preventability of the action. It would appear that tsunamis are inevitable hiccups in the natural fabric, and so they cannot be avoided. Nuclear attacks *are* avoidable, however, and so somehow appear worse, because they could have been otherwise. Yet this explanation does not capture the discrepancy in our evaluation of the events, since it is not clear that the inevitability of an event is always tied to evaluations of degrees of badness. Frankfurt has argued this time and again.[8]

Is there, then, a difference in uses of the term 'bad'? Do we mean 'bad' differently in the case of the tsunami than we do in the case of the nuclear attack? To me, the answer is clear. We mean 'bad' in an important moral sense in the latter case. The two events are not morally equivalent, which many can plainly see, even though they may have the outward appearance of being equally devastating. The fact that we can speak of the badness of a tsunami only points to the outcome of the event. It points to the loss of life, the grief, the confusion. It does not point to the responsibility of the perpetrator. The fact that we can speak with indignation about a nuclear detonation, on the other hand, certainly does point to responsibility. This observation alone is somewhat trivial, but that it is a necessity of practical reason has significant moral import.

What, then, is the process that enables us to distinguish between these events? How does the process of reasoning through the distinction here function? My suggestion is congruent with the observations of both Strawson and Habermas: that is, it is a constitutive necessity that we take up the performative or the reactive attitude as a matter of practical reason, in all interactions. This is difficult to see in epistemically obvious cases of natural events like tsunamis, because in such cases we already have ready-to-hand descriptions of the events that belie any justification that might be offered.

But try imagining the following less epistemically clear-cut case:

You are in Iceland, standing in a town at the foot of a glacier, on a placid winter day. Suddenly, the glacier lets out a thundering heave and explodes into a torrent of boiling water. What once towered over the village as a square kilometre of solid ice has mysteriously morphed into a scalding tidal wave. The town is destroyed. Thousands are killed. There is no clear explanation.

Might you not wonder what has happened here? Might you not wonder whether humans, or Gods, or alien forces, had somehow orchestrated such an event? Would your mind not leap to images of space-based lasers, or to a set of

---

[8] H. G. Frankfurt, *The Importance of What We Care About* (New York: Cambridge University Press, 1988).

carefully placed explosives? Or would your mind immediately leap to the conclusion that the Earth had just folded in on itself?

Or imagine the following:

You stumble upon a Cameroonian village known to have a population of thousands. Upon arriving, you see that all in the village have perished, seemingly at the same time. Thousands of bodies litter the streets, appearing as though they have been killed instantaneously, in the middle of a normal day. There is no evidence of a natural calamity; no fires, no brimstone, no water, no anything. Nothing. Just bodies.

Would you not suspect that such an event must have been caused by some ill-meaning outsiders? By some angry tribe or some military with biological or chemical weapons? Would you immediately think: some natural mechanism did this?

Both of the examples are natural events, though they occur only rarely, and thus do not immediately invoke an explanatory description. A jökulhlaup is a sub-glacier eruption that releases heated water in a furious torrent from where a glacier once stood. For most of the history of the world, jökulhlaups have been notoriously unpredictable, though they are feared by native Icelandic villagers. Because of their rarity, they offer few ready-to-hand objective explanations. Similarly, lake explosions, like the explosion of Kivu Lake in Cameroon in 1986 that killed 1,800 villagers, suffocate victims by instantaneously spilling tonnes of once-dissolved methane gas onto their shores. This is another obscure and rare phenomenon that does not immediately offer a ready-to-hand objective explanation.

Now clearly, both the case of the jökulhlaup and the exploding lake are destructive and awful tragedies worthy of sympathy; but they are tragedies only. They do not inspire the same degree of outrage as the attacks against the World Trade Center, as Hiroshima, as the Cambodian Killing Fields, or as the gassing of the Kurds. We don't immediately know this by evaluating the consequences. We only know this once we have worked out the specifics of how the event came to pass – of what caused the event; and whether, namely, the cause is appropriately understood solely from the perspective of the objectivating attitude; or whether we must also invoke the performative attitude.

These examples, as opposed to the nuclear/tsunami examples, add an element of mystery, an element of 'what happened here?', to the querying process. It is easy to observe and nearly instantaneously explain, for instance, that a nuclear detonation is the consequence of human activity; whereas this is much less clear when one stumbles upon a natural mystery or rare event like a lake explosion. Our natural inclination to immediately jump to the objectivating stance in the tsunami case versus the performative/reactive stance in the nuclear case masks the extent to which these stances are always already in play in our attempts to make sense of the world. That is, we don't *actively* take the intentional stance when we witness a natural catastrophe with a rough-and-ready explanation; because this stance is mostly closed and metaphorical to us. The best explanation is the one that appeals to the laws of nature. But if we stumble on something weird, like a jökulhlaup, the reaction is more subtle. We don't know how to make sense of the event – and we can be torn between both stances.

The important point is that the objectivating and performative attitudes are stances that we take at all times, at any time, as a feature of our practical reasoning. What I want to establish now, however, is that always, no matter what event has occurred, we maintain the possibility of viewing the occurrence from the perspective of either of the two stances.

Accepting this, I would like to offer a brief qualification: I am speaking mostly in this chapter about the so-called 'evils' of nature, because it is simplest to see my point if I argue along these lines. But notice that Strawson recognises reactive attitudes as working in a positive direction as well. According to Strawson, 'gratitude' is the counterpart positive emotion (as opposed to resentment) tied to the reactive attitude. If a human does something wonderful for me, perhaps saving me from a burning building, I am inclined towards gratitude and not resentment. This is different than good experiences that I view from the perspective of the objective attitude. If through some fortune (a) I stumble upon a beautiful meadow; (b) a rainstorm waters my parched crops; (c) I am rescued from drowning when a tree throws me its branch; or (d) I eat a fantastic tomato, I might be inclined, though I probably am not, to ask myself, 'Who put this here? How was this caused?', in much the same way that I ask this question about natural occurrences that I do not have a ready explanation for. The important point, as with cases of natural 'evil', is that we always maintain the possibility of seeking justification.

Communicative interaction

Habermas has introduced a potent theory of practical reasoning – a theory that insists that it is a formal requirement of our particular variety of practical reasoning that we take the second-person performative attitude when justifying actions. He argues that we can, and do, adopt the objectivating attitude when seeking to validate truth claims; but when it comes to the validation of rightness claims, we must adopt an attitude that takes seriously the will of others. In this way, he appeals to the intrapersonal standards and the justificatory apparatus unearthed by Strawson.

What I've been trying to illustrate by way of example relates to the formal requirements of practical reason, or communicative reason. It is this: that our adoption of the performative attitude is available to us and a necessary component of our reasoning through all matters, regardless of the action or event that we are assessing. We can challenge the rightness of a natural event just as much as we can challenge the rightness of an action. It may well be that rightness challenges to natural events bear little fruit, but we do, in fact, see such reactions fairly often: cancer patients feeling that their cancer is extremely unfair, for instance. Challenges of truth, rightness and truthfulness are available to us regardless of circumstance. In fact, we do, in seeking an explanation for events, always already also seek a justification for the event. Natural descriptions, causal explanations, provide for us a short-circuit out of the justificatory framework, leading us to believe that we only ever adopt a third-person objectivating attitude with regard to nature. Yet it is our peculiar human communicative predicament that we are always already in search of reasons – justificatory as well as explanatory reasons – for events and actions. Were this not so, we would always stop at evaluation of consequences; which we do not do. We evaluate – we *determine value* – by assessing events both from the objectivating and the performative attitude.

I think we can extend Strawson's and Habermas's observations to the determinism of nature, and from this develop a much clearer sense of how we should relate to nature. As I mentioned earlier, Habermas is enthusiastic about Strawson's account because it accomplishes three goals: (1) it demonstrates the necessity of the performative attitude; (2) it points to suprapersonal standards (which is to say, loosely, intrinsic values); and (3) it isolates justification as the defining feature of morality. Whether one loves or does not love nature, whether one does or does not know what nature is, one must establish one's actions as justified in order that one's actions qualify as morally permissible. Strawson seeks to disconnect the question of responsibility from the hard metaphysical question about free will. I am seeking to disconnect the question of responsibility from the hard metaphysical question about the nature of nature.

This point is as much a descriptive as it is normative. We do not respond with indignation to natural events, to natural causes. And there is a reason for this: we understand that cause in nature is determined apart from a culpable will. This is true; but this is only an after-the-fact observation, gleaned from prolonged experience with nature. In order to reason through natural causation, in order to make sense of nature, we must take it up both as an object and as an interactant.

Now, one might object that this lands us just where we began – with no further clarity on what we must value or how we should relate to nature. If there is no clarity on whether a species matters, for instance, then what has been lost? But I have a response: it is not what has been lost that makes anthropogenic species extinction so problematic, but rather how species extinction was brought about, what principles were invoked in the causation of an event. In the case of climate induced anthropogenic species extinctions, what is morally problematic is that we humans who cause the extinctions have ostensibly not taken the proper steps to ensure that the species extinction is what we are striving after. We probably have not done the scientific research required to determine the full effects of our actions, and are likely acting in such a way that we intend to bring about a particular world – say, a world of new housing developments and fancy shopping malls – without paying attention to a world bereft of species. This acting without justification, without reason, is the very *definition* of irrationality. It is a far cry from reason, autonomy, or self-governance.

If we set out to drink a glass of gin believing the liquid in our cup to be gin, when in fact it is petrol, this is a failure of reason of a particular kind. Our beliefs are not sufficient to fulfil our desires. If we set out to build a house, but in building the house, put all of the beams on upside-down, fasten them together with tomato paste, and then call our actions done, this is a failure of reason of a similar kind – we are incompetently equipped to will the means to our ends. But if we undertake to act with the expectation that our action will bring about certain circumstances, and it brings about consequences out of touch with the normative expectations of a supra-personal body – the same supra-personal body that underwrites the indignation that we feel in response to events caused by wills – which is to say, out of touch with what we have normative reasons for, quite independent of the success or failure of our actions, this is a separate kind of failure of practical reason. This is a failure of justification – a failure of practical reason that warrants moral explanation.

Environmental events, good or bad, demand explanation. It is incumbent upon us as rational beings to understand what has gone on. If the event can be established to have been anthropogenic in origin, the event also demands *justification*. Species extinctions, forest clearings, lake pollutings, seal slaughters – these events demand justification. If no justification can be found, and yet the event is determined to have emanated from the actions of rational beings, then we, as rational beings, must accept responsibility. We must accept blame. We must accept anger. We must accept shame. For what we have done is to act in such a way that is not fitting of our rational nature: that is not fitting of us. We have acted as brutes.

It is imperative, then, that we respect non-rational nature just as we respect rational nature. Why is this? Not because we rational actors must respect rationality (as many often assert), but because what it is to be rational, to be a language user, is to be respectful of reasons. Rationality entails being respectful of reasons, of justifications, and so in this respect bears a tremendous burden of proof. Rationality requires that actions be justified, and we, human language users, find ourselves under the pressure of this imperative every time we undertake to act.[9]

If we think about courses of action as courses which we choose with a justification, or for a normative reason, then I think we can understand how values emerge. When we choose a course of action – say, Route A over Route B – what we do, in part, is make a decision based on some kind of value: a want, a desire, a *pro-attitude*, as it were. We say, we are 'pro' Route A and 'nay' Route B, for instance. This pro-attitude can be understood as a 'value' because it is something that *we* value. This says nothing about the value of the thing in itself, only that we, as individual actors caught up in a network of intrapersonal normative standards, intersubjectively value one route over the other.

But given that, on the view that I am advocating, normative justifications emanate both from inside and outside an agent's subjective motivational set – which is to say, can emanate from intrapersonal or intersubjective standards – the justification provides for us a decisive indication of that which we are valuing. Value in nature is, on this picture, intersubjective.

A failure to justify

To recapitulate briefly, I have suggested that humans can and do take one of two attitudes when assessing actions – either the objective attitude or the reactive/performative attitude. I have allowed either a Strawsonian or a Habermasian view about this: that it is a fact of our psychology, or at least a fact of communicative reason, that we attribute responsibility to those who have done us harm. In order to determine who is, and who is not responsible, however, we must view all actions from the perspective of the reactive/performative attitude. We must interpret the world as a conceivable serious interactant. Acknowledging this about the natural world is key, for only once we understand non-willed causality – nature as chaotic and unchained, as law-bound but not guilt- or responsibility-bearing – can we make sense of where to place blame. Or, put differently, we can only make sense of the circumstances in which it is appropriate to hold people responsible if we have a sense of what the will is capable of effecting. Further, in taking this performative attitude to all of nature, we know where to place blame for events that precipitate in loss of life, loss of species, loss of habitat – we place the blame on free wills. Natural species extinctions, for instance, are sad and unfortunate; but only as sad and unfortunate as a predator ravaging its prey. Anthropogenic species extinctions are shameful, because the humans who cause them are blameworthy; they have acted without justification.

---

[9] Forst, *The Right to Justification*.

So this brings us back to more complicated environmental matters, including climate change and what we have done. I have thus far been skirting this question, but it should be fairly clear how the preceding discussion relates to our current predicament. Anthropogenic climate change, by definition, is a circumstance that we have brought on ourselves. Surely there are climatic shifts that have little to do with our prior actions – as climate variability is a phenomenon that predates industrial society – but many of the climate drivers now carry a strong anthropogenic signal. Carbon and methane emissions, land use practices, deforestation, etc. are all hallmarks of industrial civilisation. Certainly, much of our contribution to climate change has not come about intentionally, but rather in a disaggregated fashion.

More fundamentally, anthropogenic climate change is a consequence of billions of people acting independently by living their lives – working, driving, building, eating, breathing, consuming, excreting – and inadvertently stirring up an otherwise relatively stable state of affairs. The climatological impacts that we are discovering now are, however, a mere portion of the scope and scale of non-climatological environmental problems that we also face. Habitat destruction, ocean acidification, fisheries depletion, urban encroachment and so on, can also be easily characterised as resultant from the collective activity of billions of people going about their daily business.

When we go about our daily lives, we do so for reasons. We take our children to school, build an extension on our homes, buy a television to keep our families entertained – or, more vitally, burn a few branches to heat our food – primarily because we aim to enrich our lives and the lives of people we love. Typically our reasons for taking such actions are fairly narrow in scope. They relate to the welfare and well-being of our selves, our families, our communities and so on. As our understanding of global ecosystems has grown, however, we have begun to see that even some of the smallest, seemingly inconsequential actions can have wide-reaching cumulative impacts. All of these actions, we have learned, add up to rather substantial consequences for the Earth.

It would be false to suggest, obviously, that these negative outcomes are intentional. Indeed, nobody *intends* to destroy the planet when they purchase dinner or drive to work. But it is also false to suggest that these outcomes are accidental. Indeed, many of us are quite aware of what we are doing. This is curious indeed, since this means that the outcomes of our actions are neither accidental nor intentional. What is therefore important to note is that each of these small actions carries a normative valence. When we take a drive on a Sunday afternoon, when we pave another forest, the actions that we take *can* be and are subject to approbation or disapprobation by others, precisely because we now know that the cumulative effect of living our lives, in the ways that many of us have been living our lives, results in quite substantial shifting of the Earth's climate system. The stakes are all the more plain when we factor in harms to individual animals, loss of species diversity, or even alteration of abiotic systems. In fact, it is only in light of these considerations regarding individual animals, species diversity and alteration of abiotic systems that we can appropriately assess these otherwise everyday behaviours.

To put this a little differently, the problem of climate justice is not exactly that the Earth's climate system is shifting – though this may in fact present significant complications for life on Earth, potentially resulting in a devastating loss of resources, natural value and beauty. The problem is that when measured up against the scrutiny of outside parties, many of the actions that we take, and the reasons that support these actions, aren't of the sort that could be justified given these outcomes. We do not need to establish the moral significance of all affected corners of the Earth to understand this. All we need is to understand that the actions that we take often have impacts greater and wider than the reasons that we have for taking those actions typically account for.

Taken in this context, Robert Lewis's query about the bombing of Hiroshima takes on a sentiment perhaps just as chilling as Robert Oppenheimer's. Certainly he was concerned about the loss of life, the suffering of the injured, the profligate torching of resources and the destruction of nature, as many might assume. But more than that, he was concerned with what we had ushered in. The moment he inaugurated the nuclear age – a period during which many billions of us lived our lives in fear of unending war and nuclear annihilation. But now we face a somewhat different moment, a much longer moment ... the inauguration of the Anthropocene – a period during which many billions of us will live our lives in placid but complicit indifference to the eventual, gradual whittling away of mother nature's support systems.

The question for us, I think, is a serious one. Looking back 100 years from now, once this new moment has passed and the Anthropocene is in full swing, will we too look in the rear-view mirror and ask ourselves, as Robert Lewis once did: what have we done? If we do, if our children do, how might we explain ourselves? Could we justify what we've done? Could we explain our actions in a way that would meet the scrutiny of future generations? If we ask ourselves this question honestly, and we ask it bearing in mind all that will be affected by our mundane actions, I suspect we could arrive at an answer without too much trouble. We don't need to know how severely individual animals or species or ecosystems will be impacted by our actions. All we really need to recognise is that, up until this point, we haven't been asking this question at all.

12 Empathising with scepticism about climate change

Simon Keller

Climate change

The world's climate is changing, the change is harmful and humans are responsible.[1] Through cutting down forests, burning fossil fuels and maintaining large numbers of livestock, among other activities, humans have caused there to be greater quantities of greenhouse gases (GHGs) in the Earth's atmosphere, in turn causing temperatures to rise and weather patterns to become more erratic. In the near future, the ice caps will continue to melt; the oceans will warm, causing sea levels to rise; parts of the world in which humans presently live will become uninhabitable; hurricanes and droughts will become more common; many agricultural and other food production systems will become unsustainable, especially those relied upon by the poorest people in the poorest countries; and more non-human species will be driven to extinction.

Climate change is a singularly challenging problem technologically, but also morally and politically. The worst consequences of climate change are likely to be felt soonest by the world's most vulnerable people, but the problem is mostly caused by the activities of the world's most prosperous people. The problem of climate change crosses generations, with the acts of presently existing people likely to have their greatest effects upon people not yet born. The problem of climate change is essentially international, because it exists as a result of the accumulated actions of people in many different countries, and because changes to the climate caused by the people in one country can have their most damaging effects upon people in other countries; it is not a problem that can be dealt with through the unilateral decisions of individual states. Perhaps most significantly, climate change results from and should cause us to reconsider the culture of economic growth, ever-greater consumption and ever-greater use of energy from fossil fuels through which human progress has been

---

[1] I received helpful feedback from several people during the writing of this chapter. I would particularly like to thank Gwen Bradford, Jonathan Boston, David Coady, Garrett Cullity, Melinda Fagan, Dave Frame, Jeremy Moss, George Sher and Justin Sytsma.

understood and pursued since the Industrial Revolution. Morally and politically, the problem of climate change is a perfect storm. Only through the profoundest of changes in humanity's goals and modes of organisation could the problem be successfully confronted.

Scepticism about climate change

That, at least, is what I think. I think it is also the view of most scientists who know what they are talking about. It is certainly the view of most people I talk to. But there are many people who do not accept that view about climate change, and the numbers and influence of such people appear to be growing.[2] Some people doubt that the Earth's climate is really changing; some accept that the climate may be changing but doubt that human activity is responsible; some accept that the climate may be changing and humans may be responsible, but doubt that the problem is severe enough to require expensive or far-reaching responses. The people with these various opinions make up the class of climate change sceptics.

Climate change scepticism is a major obstacle to meaningful action on climate change. While there is widespread concern about climate change within electorates in Western democracies, the concern is not so great as to make it politically necessary for our leaders to make dealing with climate change a high priority. This is not because people are ignorant of climate change, in the sense of never having heard of it. Nor is it because people believe that climate change is occurring and has significant consequences, but do not care; it is rare to hear people say anything like, 'Yes, we are changing the Earth's climate, but the effects will be felt mostly by other people – people in poorer countries and people not yet born – so I don't think we should bother doing anything about it'. Rather, many people have the sense that the science is not settled, or that we cannot yet be sure that the problem is real and urgent, and so they do not make it a main point of pressure upon the politicians who seek their votes.

More immediately, the US is a major obstacle to real action on climate change, and part of the reason why it is an obstacle is that many influential political actors in America, including perhaps a majority of the Republican representatives in the current Congress, where the Republican Party in turn holds a majority in the House of Representatives, are climate change sceptics.[3] America is the world's most powerful country and it is responsible

---

[2] Some relevant figures are collected in E. Anderson, 'Democracy, public policy, and lay assessments of scientific testimony', *Episteme*, 8 (2011), 144–64, see p. 153.

[3] An article by Ryan Lizza in *The New Yorker* ('As the world burns', 11 October 2010) traces the efforts of some US senators to construct a bill imposing measures to address climate change. The efforts failed largely because they could not find Republican representatives who were

for a large proportion of the world's emission of GHGs. Effective international action to address climate change is impossible without American support, and American support is contingent upon Congressional support. Whatever other difficult issues are faced by efforts to confront climate change, if scepticism about climate change continues to be such a strong force within American politics, then those efforts will not succeed. If we could work out how to overcome scepticism about climate change, then we would make a significant step towards making robust action on climate change a reality.

Climate change and environmentalism

Nobody is pleased about climate change. When evidence that the world is warming and we are responsible was first made widely available, nobody took it as something to celebrate. That said, for many of us, it was not really a surprise to hear that human industrial activity is causing significant changes in the world's climate. It is a fact that fits quite smoothly with a picture of the world to which we were already committed. It is even tempting to say that the evidence about climate change stands as a vindication of that picture. Here is what I mean.

Concern about humanity's effect upon the environment predates knowledge of the nature and causes of climate change. The modern large-scale environmentalist movement predates it too. Since the eighties, at least, large numbers of us have identified as environmentalists, supporting environmental causes and feeling sympathy for green political movements. Environmentalism is marked by the conviction that 'the environment' – made up of unpolluted waterways and oceans, natural native forests, undeveloped wilderness areas and so on – is valuable and should be preserved; by the concern that our treatment of the environment will redound to the harm of humans, especially vulnerable humans in poor countries and humans yet to be born; by a rejection of the culture of unfettered economic growth and consumption and reliance upon fossil fuels; by the advocacy of a low-impact, perhaps vegetarian diet; and by support for the establishment of strong international institutions to deal with what are essentially international environmental problems.

Environmentalism does not need climate change. It focuses also on pollution, the loss of species and habitats, the destruction of places in which nature can be enjoyed, and so on. But if anything could show that the environmentalists are right, surely the evidence about climate change is it. We have been saying that humanity will be damaged by its uncaring treatment of the

sufficiently convinced of the seriousness of climate change even to enter negotiations about how the bill might look.

environment. We are not happy to hear that the climate is changing and human activity is to blame, but still – we told you so.

Accordingly, much of the philosophical treatment of climate change presses extant environmentalist themes. Climate change is linked to the exploitation of the global poor by the global rich. It is said to show that we need an entirely new approach to moral theory: one that emphasises 'green virtues' like frugality and respect for nature. It is said to show that we should eat less meat. It is said to challenge the coherence of the idea of absolute state sovereignty and to demonstrate the need for more powerful international institutions charged with taking stewardship of the global environment.[4] The news about climate change is not good news, but it does help advance the broadly environmentalist agenda, to which many of us are committed anyway.

Assessing climate change scepticism

If it is easy to view climate change from the ideological perspective of environmentalism, it is even easier to introduce ideological considerations in assessing scepticism about climate change. The basic facts about climate change have been widely known for twenty years. The scientific consensus about climate change is overwhelming. The truth is obvious. Why do so many persist in denying it?

One reason why so many people are sceptical about climate change – it is easy to point out – is that so many people are selfish and lazy. The science is out there, freely available, but people cannot be bothered reading up on it.[5] The facts about climate change are annoying. If we are meaningfully to confront climate change, then we will need to make sacrifices. We will need to pay more for petrol and use our cars less, change our diets, reduce our international travel, pay a tax on carbon and so on. For many, it is unpleasant to think that such sacrifices are necessary, and hence comforting to take refuge in the conviction that they are not.[6]

Another factor that helps to explain the prominence of climate change scepticism is the existence of a broader anti-scientific movement. The movement urges people to be distrustful of intellectuals and to put faith their own

---

[4] All of these views are pressed by several authors in the papers in S. M. Gardiner *et al.* (eds.), *Climate Change: Essential Readings* (Oxford University Press, 2010). See especially H. Shue, 'Global environment and international inequality', pp. 101–11; P. Singer, 'One atmosphere', pp. 181–99; and D. Jamieson, 'Ethics, public policy, and global warming', pp. 77–98.

[5] Anderson argues that it is quite possible for a motivated layperson to find and understand the truth about climate science, and offers several proposals for removing obstacles for their doing so. See Anderson, 'Democracy, public policy, and lay assessments of scientific testimony', especially pp. 149–53.

[6] See N. Oreskes and E. M. Conway, *Merchants of Doubt* (New York: Bloomsbury Press, 2010), pp. 266–7.

intuition, or their own religious instincts, or whatever they read on the Internet. Scepticism about climate change might be attributed partly to the same ideological orientation that lies behind the denial of evolution, and – perhaps – behind doubts about whether children should be immunised.

Finally, and most disturbingly, we can point out that scepticism about climate change is propped up by entrenched well-funded corporate and political interests whose explicit intention is to lead people astray. As was the case with tobacco, acid rain and the ozone layer, and as is often the case with pollutants and other products that threaten public health, corporations are prepared to fight against scientific results that promise to hurt their interests, whether through distorting or suppressing the scientific results, attacking the scientists responsible for the results, or publicising and funding the research of anyone approximating a scientist whose claims are less damning. And, because there are large ideological disagreements at stake, political actors who are committed to economic expansion and consumption and opposed to environmentalism have a motive to make the science look less decisive than it really is.[7]

What unites these explanations is that they take climate change scepticism to be explained by factors that are non-rational (at best). They do not call upon any considerations that could be taken, from anyone's point of view, to make scepticism about climate change more likely to be correct. Rather, they present climate change scepticism as a result of a rejection of the responsible search for truth. Climate change sceptics are made to look like people who care more about their own interests and their own dogmatic ideological commitments than they do about whether the Earth really is undergoing damaging change. To overcome climate change scepticism, it would seem, our first task is not simply to make people aware of the evidence. Our task instead is to shake people out of their selfishness and laziness and to win a political battle against corporate and ideological interests and the anti-scientific movement. The mission of overcoming climate change scepticism looks more like a war than an exercise in persuasion.

What do we know?

As easy as it is to see climate change scepticism as a result of failures of wisdom, rationality and responsibility, arguing against a real climate change sceptic is painfully difficult. I can accuse a climate change sceptic of failing to understand the basic science, but the truth is that I do not understand it either – if understanding the basic science means being able to explain it, step by step, refuting alternative hypotheses along the way, until reaching the conclusion

---

[7] For a compelling and careful exposition of this explanation, see Oreskes and Conway, *Merchants of Doubt*, esp. ch. 1.

that it is most reasonable to believe that the Earth is warming and we are responsible. I can tell a basic story about how carbon emissions lead to warming, but it is a very basic story indeed. When I am told that other explanations of warming are available, or that the purported changes to the Earth's climate are best explained as part of a natural recurring pattern, I have nothing much to say in response. I can say that the vast majority of serious scientists believe in anthropogenic climate change, and I can say that there are many scientific studies to support the thesis of anthropogenic climate change, but I cannot name many climate scientists or list the relevant studies. When someone tells me that there are better studies showing that anthropogenic climate change is not really happening, or that there is not enough evidence for the question to have been resolved, I have nowhere much to go.

The reason why I have trouble explaining why a sceptic should share my beliefs about climate change is that I come by my beliefs through trust in others' testimony. As with most complex scientific questions, I form my opinions about climate change by listening to the scientific experts.[8] That is true of almost all of us, including almost all climate change sceptics. When a believer in climate change argues with a sceptic about climate change, then, the argument is usually not about the science, directly. Instead, it is an argument about whom to trust and what the trustworthy people say.

An argument about whom we should trust is difficult to win, and in the case of climate change it can quickly turn into a political argument. You may have read something that suggests that anthropogenic climate change is a hoax, but I may offer a reason why the thing you read is not representative. I may recall a report of a study that confirms that the Earth is warming, and you may tell me that the study is likely to be biased. If the voices of climate change sceptics come mainly from outside the major institutions and journals, is that evidence that those voices are not authoritative, or is it evidence that they are being silenced by the scientific mainstream? Is the Intergovernmental Panel on Climate Change (IPCC) a straightforwardly scientific organisation, or has it become a lobbying organisation? Do scientists tend to overstate their findings out of a desire for publicity or in order to attract further funding? Is it impossible to get academic respect if you hold heretical views about climate change? If I have read *Superfreakonomics*, with its chapter on why the threat of climate change is exaggerated, does that make me better informed than you?[9] Trying to argue through these matters with a climate change sceptic is excruciating, and is certainly not – for most of us – well described as a case in which a scientifically informed party expounds the facts of the matter to a scientifically

---

[8] Anderson, 'Democracy, public policy, and lay assessments of scientific testimony', pp. 144–5; D. M. Kahan and D. Braman, 'Cultural cognition and public policy', *Yale Law and Policy Review*, 24 (2006), 147–70, see pp. 151, 155–6.

[9] S. D. Levitt and S. J. Dubner, *Superfreakonomics* (New York: William Morrow, 2009), ch. 5.

ignorant party. (For what it is worth, a recent survey suggests that supporters of the Tea Party in the US are, compared to the general population, both much more likely to be sceptical about climate change and much more likely to describe themselves as 'very informed' about climate change.[10])

To show why this may have implications for the ways in which we think about climate change scepticism, I want to offer a lengthy, fanciful, hypothetical case.

Space dust

Evidence emerges that the Earth's climate is changing. The world is becoming warmer and climate patterns more erratic, causing sea levels to rise, hurricanes and droughts to become more frequent, food production systems to be interrupted and species to be driven to extinction. The worst consequences are to be felt by poor people in poor countries and by future generations.

A number of different possible explanations of global warming are offered. Over time, however, a scientific consensus appears to build around the 'space dust hypothesis'. According to the hypothesis, the root cause of global warming is a cloud of dust that has drifted from space into the Earth's atmosphere. When it rains, the dust is brought down to the Earth's surface, where it sometimes sifts through to a layer of rock within the Earth's crust, causing a reaction that leads the rock to heat, hence warming the Earth's surface and causing overall temperatures to rise. Most of the damage is caused by rain seeping through the ground in wilderness areas, like rainforests. It appears that much of the rock that warms when brought into contact with the space dust is in these areas, and when the rain falls on treated agricultural land or runs off to rivers, the sea, or treatment plants, the warming effect is avoided.

There arises a social and political movement urging action to prevent global warming. Global warming activists urge governments to cover forested and wilderness areas with concrete, and to adopt such policies as giving large subsidies to developers, introducing new taxes on property owners who wish to leave parts of their properties undeveloped, and abolishing national parks and other protected natural areas. To prevent the planet from warming, they urge, and to prevent great harm to future generations and the world's most vulnerable humans, we need to identify the world's most naturally unspoiled areas and develop them as soon as possible.

While the science behind the space dust hypothesis is difficult for ordinary people to understand, and while the worst anticipated effects of global warming

---

[10] *The Economist*, 'American public opinion and climate change: no green tea', *The Economist Online* (8 September 2011), www.economist.com/blogs/dailychart/2011/09/american-public-opinion-and-climate-change, last accessed 10 October 2014.

are yet to be felt, for people of a certain ideological bent, the hypothesis is treated as a source of vindication. What global warming confirms, they say, is that the world is a hostile place in which we humans can survive only by bending it to our will. The story of human progress, they say, is the story of humanity taming and dominating nature. Developers and free-marketeers enthusiastically take up the cause, lecturing us on the naivety of environmentalism and the hopelessness of the idea that if we leave nature alone then somehow things will look after themselves. Only if we free humanity to innovate, develop and dominate, they say, can we save our species from destruction.

While the space dust hypothesis enjoys widespread scientific support, doubts are expressed by voices at the margins. Some scientists say that the evidence that the Earth is warming is quite flimsy, some that it is not really so clear that space dust is to blame. Some say that the warming effect of space dust will not really be counteracted by massive development of wilderness areas, some that the potential harmful effects of global warming are massively overstated. Many environmentalists speculate about the motives of the scientists who offer the space dust hypothesis, wondering whether they are trying to curry favour with the corporate interests their findings serve. Outside the scientific mainstream, people talk darkly about a conspiracy to advance the agenda of capitalists and developers using the veneer of scientific discovery.

Space dust scepticism

Faced with this scenario, I would be tempted to be a space dust sceptic. I would be suspicious of the motives of those advancing the space dust hypothesis, and very suspicious of the motives of those who want to move quickly to destroy the wilderness areas whose existence I so greatly value. At a minimum, I would urge caution. The science is difficult to understand, the promised effects of global warming seem very distant, the people I trust – environmentalists – seem unsure about the whole story, and the recommended measures are drastic. Would it not at least be worth holding out for a little longer, until we can be absolutely certain that the problem is so great, and absolutely certain that a less destructive solution cannot be found?

Space dust scepticism, of the kind just described, would not amount simply to denial. To some extent, it is rationally defensible. First, it rests upon some true suppositions. Sometimes scientists get it wrong. Sometimes scientists are moved to exaggerate the significance of their results in order to get funding or publicity. People motivated by political or selfish interests do sometimes overemphasise scientific findings that they find agreeable. Second, it rests on beliefs about the world to which I am independently committed, whether or not those beliefs turn out to be true. As an environmentalist, I am suspicious of the motivations of people who advocate greater development and destruction of wilderness areas, and I believe there is a history of corporate, political and putatively scientific parties conspiring to silence environmentalist voices and to advance an ideological agenda by stealth. Third, space dust scepticism rests partly upon a straight evaluative commitment, which is not itself under question here. Whether it is worth covering wilderness areas in concrete is a matter of weighing values, and it is quite reasonable for an environmentalist, who sees great value in keeping the wilderness in its natural state, to set a higher threshold for being convinced that such a policy is justified. The more you care about the wilderness, the greater the degree of certainty about the space dust hypothesis you will need, and the greater the harm promised by global warming will need to be, before you will agree that concreting the wilderness is a good idea.

Given my starting point as an environmentalist, scepticism about the space dust hypothesis will fit much more smoothly into my view of the world than will belief in the space dust hypothesis. My starting point may be misguided. But given that that is where I begin, it makes sense for me to be drawn to space dust scepticism. It is subjectively rational, as we might put it, for me to move from my starting point to space dust scepticism; it is a move that looks just as responsible – just as likely to get to the truth – as does any other move, from the perspective I occupy. If there is something irrational about my acceptance of space dust scepticism, then it has to do with my starting point, not with my treatment of the evidence for the space dust hypothesis or my way of drawing it within my own beliefs.

Confronting space dust scepticism

Supposing that you were a space dust sceptic, influenced by your commitment to environmentalism, what could lead you to change your mind? You could change your mind in light of a closer examination of the science, becoming convinced that the authoritative data do, in fact, support the space dust hypothesis. If the science is difficult, however, and if you are not an expert, then it is unlikely that you would come to that conclusion with any (justified) certainty. Your judgements about the science probably will (and probably should) still go by way of the testimony of experts.

You are most likely to be caused to change your mind, then, if you find experts whom you trust and who insist that the space dust hypothesis is correct. An expert you can trust will be someone whom you do not suspect of pushing an ideological agenda or being susceptible to the corrupting influence of money and power. Such an expert will be someone whom you take to share your values, or, if they have values that conflict with yours, to have successfully put their values to one side. Perhaps you could come to accept the views of experts who accept the space dust hypothesis but do not see it as the end of environmentalism. Perhaps you could be drawn to a movement that sees the prospective loss of wilderness areas as a terrible cost and is working to find alternative solutions to global warming, or that is working to minimise the amount of wilderness that needs to be sacrificed.

Given your environmentalist starting point, it will be more likely, and more reasonable, for you to accept the space dust hypothesis if the hypothesis is brought, as far as is possible, within your framework of values and your picture of the world. The greater the extent to which the hypothesis is presented as part of an anti-environmentalist agenda, the more reason you will have to be suspicious of it. Space dust scepticism can be reasonable for those who begin with a certain evaluative outlook. But, with the right kind of argument presented by the right kinds of people, perhaps it can be made reasonable for people with that outlook to give up their scepticism.

Understanding scepticism about climate change

Imagine, now, that you are the sort of person who is likely to be a sceptic about climate change, here in the real world. Suppose that you begin from the following ideological perspective: you believe that industrial activity and consumption are responsible for the progress and improved living standards of humanity over the last few hundred years; you believe that the human impulse to innovate and make money provides our best hope for continuing to lift people out of poverty; you have no particular love of 'nature' and no inclination to become vegetarian; you are suspicious of governments that gather power for themselves, and you believe that governments have a natural tendency to grow larger and levy higher taxes and exert more control wherever they can; and you are suspicious of international institutions, which you regard as undemocratic, unaccountable and likely to be dominated by persons and nations not worthy of your trust.

This starting point might be indefensible on its own terms, questions about climate change aside. Even if it is, though, it is easy to see how it could play a role in leading someone to be a climate change sceptic. Suppose that that is your starting point, and you are then confronted with the thesis of anthropogenic climate change: a thesis that you cannot fully understand and that is based on science that you are in no position to assess directly. The thesis comes to you as part of an ideologically potent package, associated with a movement that looks to impose new taxes on energy and development, to grant greater power to international organisations, to move wealth and power from the global rich to the global poor, to encourage people to eat less meat and live less consumptive lives and to alter fundamentally the economic system that has dominated in the West for the past few centuries – and that presents the news about climate change as a vindication of the environmentalist ideology. Your first thought, quite likely, will be sceptical.[11]

In deciding whether or not to trust those who advance the thesis of anthropogenic climate change, you will notice the existence of voices at the scientific margins expressing doubts about whether the thesis is correct. Much of what is said by those voices fits nicely with your pre-existing set of values and beliefs. You are told that the science is not as strong as it is said to be, that governments and universities are deliberately directing funding towards scientists who support the thesis of anthropogenic climate change, that many of those advancing the thesis are also people who are opposed to development and the free market and hence ready to abandon our prevailing economic system for any old reason, that the liberal academic establishment is deliberately shunning scientists who express doubts about the thesis, and so on. There are people available to you who claim to be experts, who doubt the reality of anthropogenic climate change, and who present their view by stating claims that you find independently plausible. You will then have some reason to trust these experts, not the climate change believers. For you, the suggestion that the scientific consensus is not authoritative will not seem extreme or fanciful. It will make sense, given the beliefs and values you already hold.

To that extent, climate change scepticism is similar to space dust scepticism, and to that extent, it can be subjectively rational.[12] The climate change sceptic might be wrong. Their ideological starting point might be misguided. But the move from their ideological starting point to their scepticism about climate change is a move that makes sense, from their own point of view. It need not be a result of laziness and selfishness, or wilful ignorance of the science, or brainwashing by corporations and right-wing politicians. It could be a result of an ordinary person doing their best to form true beliefs, within a certain social context and beginning from a certain ideological framework.

---

[11] Kahan and Braman offer evidence that views about climate change, along with many other views, tend to come as parts of distinctive ideological packages, 'Cultural cognition and public policy', pp. 149–50.

[12] My explanation of scepticism about climate change, and the suggestions I shall shortly make for addressing it, have much in common with the explanations and suggestions offered in the literature on cultural cognition: see Kahan and Braman, 'Cultural cognition and public policy'. The main point of difference is that I try to show that scepticism about climate change can emerge, in its familiar form, from a rational process. For Kahan and Braman, it is important that cultural cognition and its effects are non-rational: on their view, 'the cultural cognition of public policy can impede the rational processing of information' (p. 151). I would be happy for my argument to be interpreted as showing that the phenomena discussed in the literature on cultural cognition are due largely to differential exercises of rationality, not just to a primitive non-rational psychological mechanism.

Choosing between experts

As another way to see the nature of the predicament in which we find ourselves as ordinary non-experts trying to form a judgement about climate change, we can think about it as a particularly problematic manifestation of the general problem of choosing between experts who disagree. I have presented two stark and schematic ideologies: the environmentalist starting point of my imagined believer in climate change, and the right-wing anti-environmentalist perspective (for want of a better description) of my imagined climate change sceptic. These are not the only two possible ideological standpoints, and they are not the only two standpoints that have special salience for beliefs about climate change. Against many different backgrounds, the thesis of dangerous anthropogenic climate change is ideologically powerful, and so there are many ideological starting points that complicate the ordinary person's efforts to choose wisely between putative experts on climate change.

A guide for choosing between experts is offered by Alvin Goldman.[13] When two experts offer conflicting opinions, Goldman says, we should decide which one to trust by doing our best to evaluate the experts' respective arguments, seeking out opinions from additional experts, appraising the credentials of the experts and the judgements of authorities (like universities and professional bodies) who decide which people are the experts, asking whether the experts are likely to be biased, and looking at the experts' track records. You might use these criteria if, for example, you are a patient who has received conflicting advice from two doctors, or if you are a car-owner who has received conflicting advice from two different mechanics.

The debate about climate change has developed in such a way as to make it very difficult for a non-expert to find a non-controversial way to employ Goldman's criteria. You can try to evaluate the arguments of the different self-proclaimed experts, but the scientific arguments are complicated and difficult for a non-expert to assess, and at every point in the debate it seems that each side has a confident reply to the other. You can look for additional experts and you can ask who has the best credentials, but part of the argument of each side is to place doubt on anyone who vouches for the expertise of those on the other side; sceptics are said to be charlatans propped up by corporate interests, and believers are said to be part of a corrupt scientific establishment. You can ask about bias, but each side offers reasons to think that those on the other side are moved by self-interest or ideology, and the claims of bias on each side are difficult to confirm. You can try to assess track records, but it is difficult to know exactly what makes for a track record when it comes to explaining

---

[13] A. Goldman, 'Experts: which ones should you trust?', *Philosophy and Phenomenological Research*, 63/1 (2001), 85–110. See especially p. 93.

what will happen to the Earth's climate and why, and the facts about track records are, in any case, part of the dispute.[14] As things stand, it is much more difficult to make a sober and responsible choice about whom to trust on climate change than to make such a choice about whom to trust for medical advice, or whom to allow to fix your car.

The need for progress

I have presented my argument in the voice of a believer in dangerous anthropogenic climate change who is worried about the power of climate change scepticism. For genuine progress in confronting climate change to be achieved, I have said, it is important that we find a way to reduce the quantity and influence of scepticism in political debate. A climate change sceptic, though, should have a parallel motivation to achieve progress in the debate. As the sceptic sees it, belief in anthropogenic climate change is a costly blunder, which promises to result in misguided allocations of resources, unnecessary new taxes and regulations, and the denial of the opportunity for uninhibited industrial growth to the people in the world who most need it. For different reasons, believers and sceptics should be concerned to overcome the entrenched political division over climate change.[15] So we can all ask: how can rational progress be achieved? In closing the chapter, I want to consider three possibilities.

The triumph of the science

It is conceivable that the science could become sufficiently unambiguous and accessible for the debate to resolve itself, in one direction or the other. This, on one plausible diagnosis, is what happened in the debate over whether or not smoking causes lung cancer. Eventually, despite all the efforts of the well-funded and inventive tobacco lobby, it just became obvious to everyone that the link between smoking and lung cancer is real and scientifically established.[16] It would be nice to think that a similar resolution could be achieved in the

---

[14] Elizabeth Anderson offers a much more optimistic view about abilities of laypersons to identify the genuine experts on climate change, and the consensus the experts have reached. See 'Democracy', pp. 149–53. Anderson's account depends upon the acceptance of a hierarchy of expertise (on which PhDs from prestigious universities, for example, are more expert than others), and upon the acceptance of the integrity of academic systems of peer review, among other things. In my view, Anderson underestimates the power of the picture, offered by many climate sceptics, on which those are not indicators of expertise or truth at all.

[15] See R. W. Spencer, 'Introduction and background', in *The Great Global Warming Blunder* (New York: Encounter Books, 2010).

[16] For excellent summaries of the eventual triumph of the science in the cases of acid rain, the ozone hole and second-hand smoke, see Oreskes and Conway, *Merchants of Doubt*, chs. 3–5.

debate over climate change, but there are reasons to doubt that it will happen anytime soon.

First, the predictions made by climate scientists concern overall effects to a complex system, with consequences to be felt gradually over several decades. There are tangible, striking phenomena that can be attributed to climate change, like the melting of the polar icecaps. The tasks of showing that such phenomena are part of a predicted general pattern and showing that they are caused by the accumulation of greenhouse gases, however, is difficult to achieve to everyone's satisfaction.[17] It is much more difficult than having people realise that people who smoke are more likely to end up with lung cancer.

Second, 'climate' is a complicated concept, and 'the climate' is something that can be studied and assessed in any number of ways. The data concerning the climate include a great deal of noise, and it can be difficult to come up with clear, widely accepted, authoritative data points. For every fact about how the climate is changing in one respect, it is possible to come up with some other fact that looks like evidence that the climate is changing in the opposite respect. Putting it crudely, if it looks as though one part of the world is warming, it is always possible to come with data that make it look as though another part of the world is cooling.[18]

Third, climate science relies largely, though not solely, on modelling, and modelling is a scientific enterprise that requires good judgement and proper discretion. What counts as good judgement and proper discretion, however, is usually unclear to non-experts, and there is a heightened ability on this issue for people to accuse others of setting up their models in biased or otherwise irresponsible ways.[19]

Fourth, climate science, as a scientific field in its own right, is young. It is populated by scientists with training variously in earth sciences, meteorology, physics and chemistry, among other disciplines. There are disputes, some well motivated and some disingenuous, about what training really turns a person into an expert on climate science. One of the reasons why it is difficult to assess the claims to expertise of different self-declared climate scientists is that procedures for verifying such claims are not yet well established in the field – in comparison, at least, with the procedures established in venerable scientific disciplines like physics and biology.[20] Much climate change scepticism, indeed, is pressed by scientists who claim that they, with their training, are the true experts on climate, while those who advance the thesis of

---

[17] For some relevant discussion, see M. Hulme, *Why We Disagree about Climate Change* (Cambridge University Press, 2009), ch. 3, especially pp. 88–90.
[18] On the relationship between climate and individual weather events, see Hulme, *Why We Disagree*, pp. 5–11.
[19] *Ibid.*, pp. 56–9, 66–8.   [20] *Ibid.*, pp. 6–9, 42–68.

anthropogenic climate change in fact do not know what they are talking about, because they have been trained in the wrong disciplines.[21]

For all these reasons, I think, it will be easy for parties to the debate about climate change to keep the debate alive, or to resist acquiescence to the other side, for a good while yet.

Winning the ideological war

If the climate change sceptic and the climate change believer disagree because they have different ideological starting points, then one way to seek agreement in the debate might be to settle the underlying ideological dispute. Perhaps those who are suspicious of the scientific establishment could be convinced that their suspicions are misplaced. Perhaps those who distrust international institutions could be made to change their minds.

Some very good work has been done in trying to undermine the ideological bases of scepticism about climate change.[22] Perhaps there could also be effective parallel arguments offered from the other side, concerning the ideological framework within which belief in climate change tends to be placed. But the wars over the status of science, the value of the environment, the proper role of government and so on, are unlikely to be won, by either side, anytime soon. There is not much hope in the project of turning people into environmentalists, or of turning them away from corporatist and right-wing ideologies, and then afterwards convincing them that climate change is real.

Removing the ideology

In asking what might bring progress in a debate over the space dust hypothesis, I mentioned the significance of the source from which the science is heard and the manner in which the science is presented. You are more likely to treat someone as an expert if you lack reason to be suspicious of their ideological agenda: if their ideological commitments are the same as yours, or if you are sure that their ideological commitments are not affecting their presentation of the science. And there ought to be ways to present the evidence about climate change, as a strictly scientific matter, without making commitments as to its broader ideological consequences.

The space dust hypothesis, even if true, would not refute environmentalism. It might show that there is reason to destroy some parts of the natural

---

[21] In *Merchants of Doubt*, ch. 6, Oreskes and Conway describe the attempts of sceptical physicists to show that climate science is misunderstood by people without training in physics. Spencer's influential sceptical treatise, *The Great Global Warming Blunder*, presents itself as a meteorologist's attempt to correct the systematic mistakes made by 'climate modellers'.

[22] See Oreskes and Conway, *Merchants of Doubt*, especially chs. 6–7.

environment, but it would not show that it is wrong to value the environment or to worry about humanity's effects upon it. And climate change, by analogy, does not vindicate environmentalism, and does not show that it is wrong to value economic and industrial growth, consumption, development and state sovereignty. Taken in isolation, the thesis of dangerous anthropogenic climate change shows only that there is some cost to industrial activity, of a certain kind. It also does not show, taken in isolation, that we need a new moral theory that emphasises the green virtues, that there is something deeply wrong with capitalist consumerist culture, that it is a good thing to have strong international institutions, or that we should all be vegetarians; perhaps there are other good reasons to believe these claims, but climate change, on its own, is not enough.

Perhaps one way to take ideology out of the debate is to make it clear that the question of what to do about climate change if it exists is in principle independent of the question about whether climate change does exist. If we were not presented with the science of climate change as part of a broader package – if descriptions of what is happening to the climate were not paired with exhortations to make changes to our lifestyles, for example – then our independent assessments of elements of that broader package would not be so relevant to our decisions about whether to trust the science.[23] The process through which a person can rationally move from an ideological starting point to a certain belief about climate change might, perhaps, be interrupted.

It would be helpful for more prominence to be given to suggestions about how the thesis of anthropogenic climate change, if correct, would fit in with a greater variety of ideological perspectives. It would be good to put all options for dealing with climate change on the table, so to speak. There is nothing inconsistent, after all, about believing that if climate change is real then the best thing to do, all things considered, is just to let it happen, or to adapt rather than mitigate, or to try to confront it by using technology to further manipulate the climate. And it is always possible to agree that certain changes – new taxes, alterations to lifestyles, stronger international institutions – are necessary, but regrettable. Perhaps the measures required to combat climate change are to be taken up with reluctance, not with environmentalist enthusiasm.

The strategy of keeping the ideology out of the debate seems to me the most likely to yield genuine progress. But I have described the strategy in only the most sweeping of terms. How it could be applied at the level of political debate, in the present political climate, against the background of entrenched disagreement and mutual suspicion, I do not know.

---

[23] See the discussion of 'identity affirmation' in Kahan and Braman, 'Cultural cognition and public policy', pp. 168–70.

Conclusion

Scepticism about climate change need not be irrational. Once our predicament with regard to the science of climate change is understood, it becomes possible to see how the climate change sceptic and the climate change believer are not really so different. Much of the time, at least, climate change scepticism and climate change belief are each manifested in non-experts trying their best to form opinions about whom to trust, working within a salient ideological background and under conditions of radical informational impoverishment. The way to see how a person can be a sceptic about climate change is to try to empathise with them and their predicament, or to see how the debate looks from their point of view.

If we can understand the disagreement about climate change as arising partly through the differential use of subjectively rational strategies of belief-formation, then possibly – hopefully – we can see how progress in the debate can be achieved. And, importantly, we can see how such progress can be achieved through the engagement of people's rational capacities. People rationally form beliefs by responding to the information they have, presented in a certain way, against a certain background of pre-existing beliefs and values. Considering the various points of view from which people in our societies actually begin, how can the science of climate change be presented so as to give them good reasons to assess it on its merits, not as part of an ideological bundle? I again, do not have much of an answer, but I think it is the question we should be asking.

Bibliography

Aldred, J., 'The ethics of emissions trading', *New Political Economy*, 17/3 (2012), 339–60.
American Law Institute, *Restatement (Second) of Torts*. §§431–3 (1965).
Anderson, E., 'Democracy, public policy, and lay assessments of scientific testimony', *Episteme*, 8 (2011), 144–64.
Anscombe, G. E. M., 'War and murder', in W. Stein (ed.), *Nuclear Weapons: a Catholic Response* (London: Burns and Oates, 1961), pp. 43–62.
Armstrong, C., 'Against "permanent sovereignty" over natural resources', *Politics, Philosophy and Economics*, 14 (May 2015), 129–51.
  'Fairness, free-riding and rainforest protection', unpublished manuscript, www.academia.edu/8449971/Fairness_Free-Riding_and_Rainforest_Protection, last accessed 10 October 2014.
  'Justice and attachment to natural resources', *Journal of Political Philosophy*, 22/1 (2014), 48–65.
  'Natural resources: the demands of equality', *Journal of Social Philosophy*, 44/4 (2013), 331–47.
  'Resources, rights and global justice: a response to Kolers', *Political Studies*, 62/1 (2014), 216–22.
Arneson, R. J., 'The principle of fairness and the free-rider problem', *Ethics*, 92/4 (1982), 616–33.
Arrhenius, G., *Population Ethics* (Oxford University Press, forthcoming).
Asilomar Scientific Organizing Committee, *The Asilomar Conference Recommendations on Principles for Research into Climate Engineering Techniques* (Washington DC: Climate Institute, 2010).
Athanasiou, T. and P. Baer, *Dead Heat: Global Justice and Global Warming* (New York: Seven Stories Press, 2002).
Attfield, R., 'Mediated responsibilities, global warming, and the scope of ethics', *Journal of Social Philosophy*, 40 (2009), 225–36.
Baer, P., 'Equity, greenhouse gas emissions, and global common resources', in S. Schneider, A. Rosencranz and J. O. Niles (eds.), *Climate Change Policy: a Survey* (Washington DC: Island Press, 2002), pp. 393–408.
Barrett, S., *Environment and Statecraft: the Strategy of Environmental Treaty-making* (Oxford University Press, 2005).
Barry, B., *Democracy, Power and Justice* (Oxford: Clarendon Press, 1989).
Barry, C. and G. Øverland, 'The feasible alternatives thesis', *Politics, Philosophy and Economics*, 11/1 (2012), 97–119.

Baumert, K. A., T. Herzog and J. Pershing, 'Navigating the numbers: greenhouse gas data and international climate policy', *World Resources Institute* (2005).
Beitz, C., *Political Theory and International Relations* (Princeton University Press, 1979).
Bennett, J., *The Act Itself* (New York: Oxford University Press, 1995).
Binmore, K., *Natural Justice* (Oxford University Press, 2005).
Bipartisan Policy Centre Report, bipartisanpolicy.org/library/report/task-force-climate-remediation-research, last accessed 9 June 2014.
Blomfield, M., 'Global common resources and the just distribution of emissions shares', *Journal of Political Philosophy*, 21/3 (2013), 283–304.
Bodansky, D., 'Legitimacy', in D. Bodansky, J. Brunnee and E. Hey (eds.), *The Oxford Handbook of International Environmental Law* (Oxford University Press, 2007), pp. 711–12.
  The Art and Craft of International Environmental Law (Cambridge, MA: Harvard University Press, 2010).
  'The who, what, and wherefore of geoengineering governance', *Climatic Change*, 121 (2013), 539–51.
Boucher, O., P. M. Forster, N. Gruber, M. Ha-Duong, M. G. Lawrence *et al.*, 'Rethinking climate engineering categorization in the context of climate change mitigation and adaptation', *WIREs Climate Change*, 5 (2014).
Bradley, B., 'Doing away with harm', *Philosophy and Phenomenological Research*, 85/2 (2012), 390–412.
Bratman, M. E., 'Shared cooperative activity', *Philosophical Review*, 101 (1992), 327–40.
Brennan G. and J. M. Buchanan, *The Reason of Rules: Constitutional Political Economy* (Indianapolis, IN: Library of Economics and Liberty, 2000), www.econlib.org/library/Buchanan/buchCv10.html, accessed 21 May 2014.
Brighouse, H. and M. Fleurbaey, 'Democracy and proportionality', *Journal of Political Philosophy*, 18/2 (2010), 137–55.
Broome, J., *Climate Matters: Ethics in a Warming World* (New York: W. W. Norton and Company, 2012).
  'The public and private morality of climate change', *The Tanner Lectures on Human Values*, University of Michigan, 16 March 2012.
  Weighing Goods (Oxford: Blackwell, 1991).
  Weighing Lives (Oxford University Press, 2004).
Bunzl, M., 'Causal overdetermination', *Journal of Philosophy*, 76/3 (1979), 134–50.
Butt, D., 'A doctrine quite new and altogether untenable: defending the beneficiary pays principle', *Journal of Applied Philosophy*, 31/4 (2014), 336–48.
Caldeira, K. and D. W. Keith, 'The need for climate engineering research', *Issues in Science and Technology*, 27 (2010), 57–62.
Caney, S., 'Climate change and the duties of the advantaged', *Critical Review of International Social and Political Philosophy*, 13/1 (2010), 203–28.
  'Climate change, human rights, and moral thresholds', in S. M. Gardiner, S. Caney and H. Shue (eds.), *Climate Ethics: Essential Readings* (Oxford University Press, 2010), pp. 163–77.
  'Cosmopolitan justice, responsibility, and global climate change', *Leiden Journal of International Law*, 18/4 (2005), 747–75.

'Environmental degradation, reparations, and the moral significance of history', *The Journal of Social Philosophy*, 37/3 (2006), 464–82.

'Just emissions', *Philosophy and Public Affairs*, 40/4 (2013), 255–300.

'Two kinds of climate justice: avoiding harm and sharing burdens', *Journal of Political Philosophy*, 21/4 (2014), 125–49.

Caney, S. and C. Hepburn, 'Carbon trading: unethical, unjust and ineffective?', *Royal Institute of Philosophy, Supplement* 69 (2011), 201–34.

Cardwell, M., *Milk Quotas, European Community and United Kingdom Law* (Oxford: Clarendon Press, 1996).

Christiano, T. 'Equality, fairness and agreements', *Journal of Social Philosophy: Special Issue on New Directions in Egalitarianism*, 44/4 (2013), 370–91.

'Rational deliberation among citizens and experts', in J. Mansbridge and J. Parkinson (eds.), *Deliberative Systems: Deliberative Democracy at the Large Scale* (Cambridge University Press, 2012).

'The legitimacy of international institutions', in A. Marmor (ed.), *The Routledge Companion to the Philosophy of Law* (New York: Routledge, 2012).

The Rule of the Many (Boulder, CO: Westview Press, 1996).

Christiano, T. and W. Braynen, 'Inequality, injustice and leveling down', *Ratio*, 21 (2008), 392–420.

Cicerone, R. J., 'Geoengineering: encouraging research and overseeing implementation', *Climatic Change* 77/3 (2006), 221–6.

Cohen, G. A., *If You're an Egalitarian, How Come you're So Rich?* (Cambridge, MA: Harvard University Press, 2000).

'Luck and equality', *Philosophy and Phenomenological Research*, 72 (2006), 439–46.

'On the currency of egalitarian justice', *Ethics*, 99 (1989), 906–44.

Rescuing Justice and Equality (Cambridge, MA: Harvard University Press, 2008).

Cohen, S. and S. Spacapan, 'The social psychology of noise', in D. M. Jones and A. J. Chapman (eds.), *Noise and Society* (Chichester: John Wiley, 1984), pp. 221–45.

Conor, S. 'Sun sets on sceptics' case against climate change', *The Independent*, Monday, 14 December 2009, www.independent.co.uk/environment/climate-change/sun-sets-on-sceptics-case-against-climate-change-1839875.html, last accessed 10 October 2014.

Convention on Biological Diversity, www.cbd.int/climate/geoengineering/default.shtml, last accessed 28 May 2014.

Convention on Environmental Modification, www.un-documents.net/enmod.htm, last accessed 28 May 2014.

Cripps, E. 'Climate change, collective harm and legitimate coercion', *Critical Review of International Social and Political Philosophy*, 14/2 (2011), 171–93.

Cullity, G., 'Moral free riding', *Philosophy and Public Affairs*, 24/1 (1995), 3–34.

'Public goods and fairness', *Australasian Journal of Philosophy*, 86 (2008), 1–21.

'The moral, the personal, and the political', in I. Primoratz (ed.), *Politics and Morality* (Basingstoke: Palgrave Macmillan, 2007), pp. 54–75.

Darwall, S. L., *The Second-person Standpoint: Morality, Respect, and Accountability* (Cambridge, MA: Harvard University Press, 2006),

de la Croix, D. and A. Gosseries, 'Population policy through tradable procreation entitlements', *International Economic Review*, 50 (2009), 507–42.

Dobson, A., *Citizenship and Environment* (Oxford University Press, 2003).

Bibliography

Ellerman, D., P. L. Joskow, R. Schmalensee, J.-P. Montero and E. M. Bailey, *Markets for Clean Air: the US Acid Rain Program* (New York: Cambridge University Press, 2000).

Ellis, E. C., 'Anthropogenic transformation of the terrestrial biosphere', *Philosophical Transactions of the Royal Society A: Mathematical, Physical and Engineering Sciences*, 369 (2011), 1010–35.

ETC, 'IPCC and geoengineering: the bitter pill is also a poison pill', ETC News Release, 16 April 2014, www.etcgroup.org/content/ipcc-and-geoengineering-bitter-pill-also-poison-pill, last accessed 20 May 2014.

Feinberg, J., *Harm to Others: the Moral Limits of the Criminal Law* (Oxford University Press, 1984).

Fleurbaey, M., 'Justice et climat: alliance ou tension?', *Raison Publique*, April 2010.

Forst, R., 'The justification of human rights and the basic right to justification: a reflexive approach', *Ethics*, 120 (2010), 711–40.

The Right to Justification: Elements of a Constructivist Theory of Justice, J. Flynn (trans.), (New York: Columbia University Press, 2012).

Frankfurt, H. G., *The Importance of What We Care About* (New York: Cambridge University Press, 1988).

Gardiner, S. M., *A Perfect Moral Storm: the Ethical Tragedy of Climate Change* (Oxford University Press, 2011).

'Ethics and global climate change', in S. M. Gardiner, S. Caney and H. Shue (eds.), *Climate Ethics: Essential Readings* (Oxford University Press, 2010), pp. 3–35.

'Some early ethics of geoengineering the climate: a commentary on the values of the Royal Society report', *Environmental Values*, 20/2 (May 2011), 163–88.

Gardiner, S. M, S. Caney and H. Shue (eds.), *Climate Change: Essential Readings* (Oxford University Press, 2010).

German Advisory Council on Global Change (WBGU), *Solving the Climate Dilemma: the Budget Approach* (Berlin, 2009), www.wbgu.de/en/publications/special-reports/special-report-2009, last accessed 10 October 2014.

Gilbert, M., *Living Together: Rationality, Sociality, and Obligation* (Lanham, MD: Rowman and Littlefield, 1996).

Global Humanitarian Forum, *The Anatomy of a Silent Crisis: Climate Change Human Impact Report* (2009), www.ghf-ge.org/human-impact-report.pdf, last accessed 10 October 2014.

Glover, J., *Causing Death and Saving Lives* (Harmondsworth: Penguin, 1977).

Goldman, A. I., 'Experts: which ones should you trust?' *Philosophy and Phenomenological Research*, 63/1 (2001), 85–110.

'Why citizens should vote: a causal responsibility approach', *Social Philosophy and Policy*, 2 (1999), 201–17.

Goodin, R., 'Selling environmental indulgences', in J. Dryzek and D. Schlosberg (eds.), *Debating the Earth* (Oxford University Press, 1994), pp. 237–54.

Goodin, R. and C. Barry, 'Benefitting from the wrong doing of others', *Journal of Applied Philosophy*, 31/2 (2014), 363–76.

Goodin, R. E. and J. Dryzek, 'Risk sharing and justice: the motivational foundations of the post-war welfare state', in R. Goodin and J. LeGrand (eds.), *Not Only the Poor: the Middle Classes and the Welfare State* (London: Allen and Unwin, 1987), pp. 37–73.

Gosseries, A., 'Cosmopolitan luck egalitarianism and climate change', *Canadian Journal of Philosophy*, suppl. 31 (2007), 279–309.
  'Historical emissions and free-riding', *Ethical Perspectives*, 11/1 (2004), 36–60.
Gosseries, A. and V. Van Steenberghe, 'Pourquoi les marches de permis de polluter? Les enjeux économiques et éthiques de Kyoto', *Regards Économiques*, 21 (2004), 1–14.
Grotius, H., *On the Law of War and Peace*, student edn, S. Neff (ed.), (Cambridge University Press, 2012).
Habermas, J., *Between Facts and Norms* (Cambridge, MA: MIT Press, 1994).
  'Discourse ethics', in C. Lenhardt and S. W Nicholson (eds.), *Moral Consciousness and Communicative Action* (Cambridge, MA: MIT Press, 1991).
  Truth and Justification (Cambridge, MA: MIT Press, 2003).
Hale, B., 'Can we remediate wrongs?', in A. Hiller, R. Ilea and L. Kahn (eds.), *Consequentialism and Environmental Ethics* (New York, NY: Routledge, 2013).
  'Getting the bad out: remediation technologies and respect for others', in J. K Cambell (eds.), *The Environment: Topics in Contemporary Philosophy*, vol. 9. (Cambridge, MA: MIT Press, 2012).
  'Moral considerability: deontological, not metaphysical', *Ethics and the Environment*, 16/2 (2011), 37–62.
  'Polluting and unpolluting', in M. Boylan (ed.), *Environmental Ethics*, 2nd edition (Hoboken, NJ: Wiley-Blackwell, 2013).
  'Technology, the environment, and the moral considerability of artifacts', in D. E. Selinger, J. K. B. Olsen and S. Riis (eds.), *New Waves in Philosophy of Technology* (London: Ashgate, 2008), pp. 216–40.
Hale, B., A. Hermans and A. Lee, 'Adaptation, reparation, and the baseline problem', in M. Boykoff and S. Moser (eds.), *Toward Successful Adaptation: Linking Science and Practice in Managing Climate Change Impacts* (London and New York: Routledge, 2013), pp. 67–80.
Hale, B., A. Lee and A. Hermans, 'Clowning around with conservation: adaptation, reparation, and the new substitution problem', *Environmental Values*, 23 (2014) 181–98.
Hall, N., 'Non-locality on the cheap? A new problem for counterfactual analyses of causation', *Noûs*, 36/2 (2002), 276–94.
  'Two concepts of causation', in J. Collins, N. Hall and L. A. Paul (eds.), *Causation and Counterfactuals* (Cambridge, MA: MIT Press, 2004), pp. 225–76.
Halliday, D., 'Review essay of *Justice, Institutions and Luck*', *Utilitas*, 25/1 (2013), 121–32.
Hansen, J., M. Sato, P. Kharecha, D. Beerling, R. Berner *et al.*, 'Target atmospheric CO_2: where should humanity aim?', *Open Atmospheric Science Journal*, 2 (2008), 217–31.
Hanser, M., 'The metaphysics of harm', *Philosophy and Phenomenological Research*, 77/2 (2008), 421–50.
Hare, C., 'Voices from another world: must we respect the interests of people who do not, and will never, exist?', *Ethics*, 117 (2007), 498–523.
Harman, E., 'Harming as causing harm', in M. A. Roberts and D. Wasserman (eds.), *Harming Future Persons* (Dordrecht: Springer, 2009), pp. 137–54.
Harsanyi, J. C., 'Can the maximin principle serve as a basis for morality? A critique of John Rawls's theory', *American Political Science Review*, 69 (1975), 594–606.

Hart, H. L. A. and T. Honoré, *Causation in the Law*, 2nd edition (Oxford: Clarendon Press, 1985).
Hathaway, J. and A. Neve, 'Making international refugee law relevant again: a proposal for a collectivized and solution-oriented protection', *Harvard Human Rights Journal*, 10 (1997).
Held, V. 'Can a random collection of individuals be morally responsible?', *Journal of Philosophy*, 67 (1970), 471–81.
Helm, D., 'Climate change policy: why has so little been achieved?' in D. Helm and C. Hepburn (eds.), *The Economics and Politics of Climate Change* (Oxford University Press, 2011), pp. 9–35.
Heyward, C., 'Situating and abandoning geoengineering: a typology of five responses to dangerous climate change', *Political Science and Politics*, 46/1 (2013), 23–7.
Higgs, E., 'Changing nature: novel ecosystems, intervention, and knowing when to step back sustainability science', in M. P. Weinstein and R. E. Turner (eds.), *Sustainability Science: the Emerging Paradigm and the Urban Environment* (New York: Springer, 2012).
Hinkle Charitable Foundation, *How do we Contribute Individually to Global Warming?*, www.thehcf.org/emaila5.html, last accessed 27 October 2014.
Hirose, I., *Egalitarianism* (London: Routledge, 2014).
Hobbes, T., *The Leviathan*, J. C. A Gaskin (ed.), (Oxford University Press, 1996 [1651]).
Hobbs, R. J., E. Higgs and J. A Harris, 'Novel ecosystems: implications for conservation and restoration', *Trends in Ecology and Evolution*, 24 (2009), 599–605.
Hoekman, B. and M. Kosteki, *The Political Economy of the World Trading System: the WTO and Beyond*, 3rd edition (Oxford University Press, 2009).
Höhne, N., J. Kejun, J. Rogelj, L. Segafredo, R. S. da Motta and P. R. Shukla, *The Emissions Gap Report 2012: a UNEP Synthesis Report* (Nairobi: United Nations Environment Programme (UNEP), place: publisher, 2012), www.unep.org/publications/ebooks/emissionsgap2012.
Holtug, N., *Persons, Interests, and Justice* (Oxford University Press, 2010).
Honore, A. M., 'Necessary and sufficient conditions in tort law', in D. G Owen (ed.), *Philosophical Foundations of Tort Law* (Oxford: Clarendon Press, 1995), pp. 363–85.
Hulme, M., *Why We Disagree about Climate Change* (Cambridge University Press, 2009).
Intergovernmental Panel on Climate Change (IPCC), *Climate Change 2001: Synthesis Report. A Contribution of Working Groups I, II, and III to the Third Assessment Report of the IPCC*. R. T. Watson and the core writing team (eds.), (Cambridge, UK and New York: Cambridge University Press, 2001).
  Climate Change 2007: Impacts, Adaptation and Vulnerability. Contribution of Working Group II to the Fourth Assessment Report of the Intergovernmental Panel on Climate Change, M. L. Parry et al. (eds.), (Cambridge, UK and New York: Cambridge University Press, 2007).
  Climate Change 2013: the Physical Science Basis. Contribution of Working Group I to the Firth Assessment Report of the Intergovernmental Panel on Climate Change, T. F. Stocker, D. Qin, G.-K. Plattner, M. Tignor, S. K. Allen et al. (eds.), (Cambridge, UK and New York, USA: Cambridge University Press, 2013), pp. 1535.
  Climate Change 2014: Impacts, Adaptation and Vulnerability, www.ipcc.ch/report/ar5/wg2, last accessed 10 October 2014.

Managing the Risks of Extreme Events and Disasters to Advance Climate Change Adaptation: Summary for Policymakers', *A Special Report of Working Groups I and II of the IPCC*, C. B. Field et al. (eds.), (Cambridge, UK and New York: Cambridge University Press, 2012), pp. 1–19.
Working Group III, *Summary for Policymakers*, 25.
Irvine, P. J., A. Ridgwell and D. J. Lunt, 'Assessing the regional disparities in geoengineering impacts', *Geophysical Research Letters*, 37/18 (2010).
'Climatic effects of surface albedo geoengineering', *Journal of Geophysical Research*, 116/D24 (2011).
Jackson, F. 'Which effects?', In J. Dancy (ed.), *Reading Parfit* (Oxford: Blackwell, 1997), pp. 42–53.
Jaeger, C. and J. Jaeger, 'Three views of two degrees', *Regional Environmental Change*, 11 (2011), 815–26.
Jamieson, D., 'Ethics and intentional climate change', *Climatic Change*, 33 (1996), 331–2.
'Ethics, public policy, and global warming', in S. M Gardiner, S. Caney and H. Shue (eds.), *Climate Ethics: Essential Readings* (Oxford University Press, 2010), pp. 77–98.
Jamieson, D., *Reason in a Dark Time* (New York: Oxford University Press, 2014).
Jeffrey, R., *The Logic of Decision*, 2nd edition (University of Chicago Press, 1983).
Johnson, B., 'Ethical obligations in a tragedy of the commons', *Environmental Values*, 12 (2003), 271–87.
Jones, A., J. Haywood and O. Boucher, 'Climate impacts of geoengineering marine stratocumulus clouds', *Journal of Geophysical Research*, 114/D10 (2009).
Kagan, S., 'Do I make a difference?', *Philosophy and Public Affairs*, 39 (2011), 105–41.
Kahan, D. M. and D. Braman, 'Cultural cognition and public policy', *Yale Law and Policy Review*, 24 (2006), 147–70.
Kanowski, P., C. McDermott and B. Cashore, 'Implementing REDD+: lessons from analysis of forest governance', *Environmental Science and Policy*, 14/2 (2011), 111–17.
Kant, I. *The Metaphysics of Morals*, M. Gregor (trans), (Cambridge University Press, 1991 [1797]).
Keith, D. W., E. Parson and M. G. Morgan, 'Research on global sun block needed now', *Nature*, 463 (2010), 426–7.
Keohane, R. and D. Victor, 'The regime complex for climate change', *Perspectives on Politics*, 9/1 (2011), 7–23.
Keohane, R. and J. S. Nye Jr., 'The club model of multilateral cooperation and problems of democratic legitimacy', in R. Keohane (ed.), *Power and Governance in a Partially Globalized World* (London: Routledge Publishers, 2002), pp. 219–44.
Keohane, R., S. Macedo and A. Moravscik, 'Democracy enhancing multilateralism', *International Organization*, 63/1 (2009), 1–31.
Killoren, D. and Williams, B. 'Group agency and overdetermination', *Ethical Theory and Moral Practice*, 16 (2013), 295–307.
Kingsbury, B., N. Krisch and R. Stewart, 'The emergence of global administrative law', *Law and Contemporary Problems*, 68/3 (2005), 15–62.
Kolers, A., 'Justice, territory and natural resources', *Political Studies*, 60/2 (2012), 269–86.

Kolstad, C. D., 'Piercing the veil of uncertainty in transboundary pollution agreements', *Environmental and Resource Economics*, 31 (2005), 21–34.
Kolstad C. D. and A. Ulph, 'Uncertainty, learning and heterogeneity in international environmental agreements', *Environmental and Resource Economics*, 50/3 (2011), 389–403.
Lane, L., 'Climate engineering and international law: what is in the national interest?' *Proceedings of the Annual Meeting (American Society of International Law)*, 105 (2011), 525–8.
Larson, A., 'Forest tenure reform in the age of climate change: lessons for REDD+', *Global Environmental Change*, 21/2 (2011), 540–49.
Lawford-Smith, H., 'The feasibility of collectives' actions', *Australasian Journal of Philosophy*, 90 (2012), 53–67.
Levitt S. D. and S. J. Dubner, *Superfreakonomics* (New York: William Morrow, 2009).
Lewis, D., 'Causation as influence', *Journal of Philosophy*, 97/4 (2000), 182–97.
Lippert-Rasmussen, K., 'Hurley on egalitarianism and the luck-neutralizing aim', *Politics, Philosophy, and Economics*, 4 (2005), 249–65.
  'Inequality, incentives, and the interpersonal test', *Ratio*, 21 (2008), 421–39.
  'Luck-egalitarianism: faults and collective choice', *Economics and Philosophy*, 27 (2011), 151–73.
  Luck Egalitarianism (London: Bloomsbury, 2015).
Lizza, R., 'As the world burns', *The New Yorker* (11 October 2010).
Locke, J., *Two Treatises of Government*, P. Laslett (ed.), (Cambridge University Press, 1960).
London Convention and Protocol, imo.org/OurWork/Environment/LCLP/Pages/default.aspx, last accessed 28 May 2014.
Lunt, D. J., A. Ridgwell, P. J. Valdes and A. Seale, '"Sunshade world": a fully coupled GCM evaluation of the climatic impacts of geoengineering', *Geophysical Research Letters*, 35/12 (2008).
MacCracken, M. 'Geoengineering: worthy of cautious evaluation?' *Climatic Change*, 77 (2006), 235–43.
Mackie, J. L., *Cement of the Universe*, 2nd edition (Oxford University Press, 1980).
Meinshausen, M., N. Meinshausen, W. Hare, S. C. B. Raper, K. Frieler *et al.*, 'Greenhouse-gas emission targets for limiting global warming to 2 degrees C', *Nature*, 458 (2009), 1158–62.
Meyer, A., *Contraction and Convergence: the Global Solution to Climate Change* (Dartington, UK: Green Books, 2000).
Meyer, L. H., 'Compensating wrongless historical emissions of greenhouse gases', *Ethical Perspectives*, 11/1 (2004), 20–35.
Meyer, L. and P. Sanklecha, 'Individual expectation and climate justice', *Analyze and Kritik*, 2 (2011), 449–71.
Mill, J. S., *On Liberty* (Buffalo, NY: Prometheus Books, 1986).
Miller, D., 'Global justice and climate change: how should responsibilities be distributed?', *The Tanner Lectures on Human Values*, Tsinghua University, Beijing, 24–5 March (2008), http://tannerlectures.utah.edu/_documents/a-to-z/m/Miller_08.pdf, last accessed 10 October 2014.
  National Responsibility and Global Justice (Oxford University Press, 2007).
  'Territorial rights: concept and justification', *Political Studies*, 60/2 (2012), 252–68.

Moellendorf, D., 'Treaty norms and climate mitigation', *Ethics and International Affairs*, 23/3 (2009), 247–65.
Moore, M., 'For what must we pay? Causation and counterfactual baselines', *San Diego Law Review*, 40 (2003), 1181–271.
Morgan-Knapp, C. and C. Goodman, 'Consequentialism, climate harm and individual obligations', *Ethical Theory and Moral Practice* (2014).
Morrow, D. R., R. E. Kopp and M. Oppenheimer, 'Toward ethical norms and institutions for climate engineering research', *Environmental Research Letters*, 4 (2009).
Moss, J., *Reassessing Egalitarianism* (London: Palgrave McMillan, 2014).
Murphy, L., 'Institutions and the demands of justice', *Philosophy and Public Affairs*, 27/4 (1999), 251–91.
  'The demands of beneficence', *Philosophy and Public Affairs*, 22/4 (1993), 267–92.
Na, S. and H. S. Shin, 'International environmental agreements under uncertainty', *Oxford Economic Papers*, 50 (1998), 173–85.
Nagel, T., 'The problem of global justice', *Philosophy and Public Affairs*, 33 (2005), 113–47.
Narain U. and K. van't Veld, 'The clean development mechanism's low-hanging fruit problem: when might it arise, and how might it be solved?', *Environmental Resource Economics*, 40 (2008), 445–65.
Narveson, J., 'Moral problems of population', *The Monist*, 57 (1973), 62–86.
Neumayer, E., 'In defense of historical accountability for greenhouse gas emissions', *Ecological Economics*, 33 (2000), 185–92.
Nine, C., *Global Justice and Territory* (Oxford University Press, 2012).
Nolt, J., 'How harmful are the average American's greenhouse gas emissions?', *Ethics, Policy and Environment*, 14/1 (2011), 3–10.
Nordhaus, W., *A Question of Balance: Weighing Options on Global Warming Policies* (Yale University Press, 2008).
  'A review of the *Stern Review* on the economics of climate change', *Journal of Economic Literature*, 45 (2007), 686–702.
North Pacific Fur Seal Treaty, Article III (1911).
Nozick, R., *Anarchy, State and Utopia* (New York: Basic Books, 1974).
Oreskes, N. and E. M. Conway, *Merchants of Doubt* (New York: Bloomsbury Press, 2010).
Page, E., 'Cashing in on climate change: political theory and global emissions trading', *Critical Review of International Social and Political Philosophy*, 14/2 (2011), 259–79.
  'Climatic justice and the fair distribution of atmospheric burdens: a conjunctive account', *Monist*, 94/3 (2011), 412–32.
  'Distributing the burdens of climate change', *Environmental Politics*, 17 (2008), 556–75.
  'Give it up for climate change: a defence of the beneficiary pays principle', *International Theory*, 4 (2012), 300–30.
  'Intergenerational justice and climate change', *Political Theory*, 47 (1999), 53–66.
  'Intergenerational justice of what: welfare, resources or capabilities?', *Environmental Politics*, 16 (2007), 453–69.
  'Equality and priority', in A. Mason (ed.), *Ideals of Equality* (Oxford: Blackwell, 1998), pp. 1–20.
  Reasons and Persons (Oxford: Clarendon Press, 1983).

Parson, E. A., Parfit, D. and L. N. Ernst, 'International governance of climate engineering', *Theoretical Inquiries in Law*, 14 (2013), 307–38.
Persson, I., 'Why leveling down could be worse for prioritarianism than for egalitarianism', *Ethical Theory and Moral Practice*, 11 (2008), 295–303.
Phelps, J., E. Webb and A. Agrawal, 'Does REDD+ threaten to recentralize forest governance?', *Science*, 328/5976 (2010), 312–31.
Pogge, T., 'An egalitarian law of peoples', *Philosophy and Public Affairs*, 23/3 (1994), 195–224.
  'The categorical imperative', in P. Guyer (ed.), *Kant's "Groundwork of the Metaphysics of Morals": Critical Essays* (Lanham, MD: Rowman and Littlefield, 1998), pp. 189–213.
Posner E. A. and D. Weisbach, *Climate Change Justice* (Princeton University Press, 2010).
Preston, C. J., 'Ethics and geoengineering: reviewing the moral issues raised by solar radiation management and carbon dioxide removal', *WIREs Climate Change*, 4 (2013), 23–37.
Rakowski, E., *Equal Justice* (Oxford: Clarendon Press, 1991).
Ramsey, F., 'Truth and probability', in D. H. Mellor (ed.), *Foundations: Essays in Philosophy, Logic, Mathematics and Economics* (London: Routledge and Kegan Paul, 1978), pp. 58–100.
Rawls, J., *A Theory of Justice* (Cambridge, MA: Harvard University Press, 1971).
  A Theory of Justice: Revised Edition (Oxford University Press, 1999).
  Justice as Fairness (Cambridge, MA: Harvard University Press, 2001).
  Political Liberalism (New York: Columbia University Press, 1993).
  The Law of Peoples (Cambridge, MA: Harvard University Press, 1999).
Rayner, S., 'To know or not to know? A note on ignorance as a rhetorical resource in geoengineering debates', *Climate Geoengineering Governance Working Paper Series*, 10 (2014).
Rayner, S., C. Heyward, T. Kruger, N. Pidgeon, C. Redgwell and J. Savulescu, 'The Oxford Principles', *Climatic Change*, 121 (2013), 499–512.
Raz, J., *Practical Reason and Norms*, 2nd edition (Oxford: Clarendon Press, 1990).
Ricke, K. L., M. G. Morgan and M. R. Allen, 'Regional climate response to solar-radiation management', *Nature Geoscience*, 3 (2010), 537–41.
Risse, M., *On Global Justice* (Princeton University Press, 2012).
Robock, A., L. Oman and G. L. Stenchikov, 'Regional climate responses to geoengineering with tropical and arctic SO2 injections', *Journal of Geophysical Research*, 113/D16 (2008).
Robock, A., M. Bunzl, B. Kravitz and G. L. Stenchikov, 'A test for geoengineering?', *Science*, 327/5965 (2010), 530–1.
Sandberg, J., 'My emissions make no difference: climate change and the argument from inconsequentialism', *Environmental Ethics*, 33 (2011), 229–48.
Sandel, M., 'It's immoral to buy the right to pollute', in M. Sandel, *Public Philosophy: Essays on Morality in Politics* (Cambridge, MA: Harvard University Press, 2005), pp. 93–6.
Satz, D., *Why Some Things Should Not Be for Sale: the Moral Limits of Markets* (Oxford University Press, 2010).
Savage, L., *The Foundations of Statistics* (New York: John Wiley and Sons, 1954).

Scanlon, T. M., *What We Owe to Each Other* (Cambridge, MA: Harvard University Press, 1998).
Schaffer, J., 'Overdetermining causes', *Philosophical Studies*, 114 (2003), 23–45.
Scheffler, S., *Boundaries and Allegiances: the Problems of Justice and Responsibility in Liberal Thought* (Oxford University Press, 2001).
Schroeder, D. and T. Pogge, 'Justice and the convention on biological diversity', *Ethics and International Affairs*, 23/3 (2009), 267–80.
Schuck, P., 'Refugee burden-sharing: a modest proposal', *Yale Journal of International Law*, 22 (1997), 243–97.
Segall, S., 'Why egalitarians should not care about equality', *Ethical Theory and Moral Practice*, 15 (2012), 507–19.
Shepherd, J., K. Caldeira, P. Cox, J. Haigh, D. Keith et al., *Geoengineering the Climate: Science, Governance and Uncertainty* (London: Royal Society, 2009).
Shiffrin, S., 'Wrongful life, procreative responsibility, and the significance of harm', *Legal Theory*, 5 (1999), 117–48.
Shue, H. 'Exporting hazards', *Ethics*, 91/4 (1981), 579–606.
  'Global environment and international inequality', in S. M Gardiner, S. Caney and H. Shue (eds.), *Climate Ethics: Essential Readings* (Oxford University Press, 2010), pp. 101–11.
  'Global environment and international inequality', *International Affairs*, 75/3 (2003), 531–45.
  'Subsistence emissions and luxury emissions', *Law and Policy*, 15/1 (1993), 39–59.
Shui, B. and R. Harris, 'The role of CO_2 embodiment in US–China trade', *Energy Policy*, 34/18 (2006) 4063–8.
Simmons, A. J., *Moral Principles and Political Obligations* (Princeton University Press, 1979).
Singer, P., 'Famine, affluence and morality', *Philosophy and Public Affairs*, 1/1 (1972), 229–43.
  'One atmosphere', in S. M Gardiner, S. Caney and H. Shue (eds.), *Climate Ethics: Essential Readings* (Oxford University Press, 2010), pp. 181–99.
  One World (Melbourne: Text Publishing, 2002), p. 39.
  One World: the Ethics of Globalization, 2nd edition (New Haven: Nota Bene Press, 2004).
  Practical Ethics, 3rd edition (Cambridge University Press, 2011).
Sinnott-Armstrong, W., '"It's not *my* fault": global warming and individual moral obligations', in W. Sinnott-Armstrong and R. B. Howarth (eds.), *Perspectives on Climate Change: Science, Economics, Politics, Ethics*. Advances in the Economics of Environmental Resources (Amsterdam: Elsevier, 2005), vol. V, pp. 285–307.
Spash, C., 'The brave new world of carbon trading', *New Political Economy*, 15/2 (2010) 169–95.
Spencer, R. W., *The Great Global Warming Blunder* (New York: Encounter Books, 2010).
Stapleton, J., 'Choosing what we mean by "causation" in law', *Missouri Law Review*, 73 (2008), 433–80.
  'Legal cause: cause-in-fact and the scope of liability for consequences', *Vanderbilt Law Review*, 54 (2001), 941–1009.
Steiner, H., *An Essay on Rights* (Oxford: Blackwell, 1994).

Stern, N., *Stern Review of the Economics of Climate Change* (2006), http://webarchive.nationalarchives.gov.uk/+/http://www.hm-treasury.gov.uk/sternreview_summary.htm, last accessed 10 October 2014.
  The Economics of Climate Change: the Stern Review (Cambridge University Press, 2007).
  'The economics of climate change', in S. M. Gardiner, S. Caney and H. Shue (eds.), *Climate Ethics: Essential Readings* (Oxford University Press, 2010), pp. 39–86.
Stilz, A., 'Nations, states and territory', *Ethics*, 121/3 (2011), 575–601.
Stratton-Lake, P., 'Introduction', in W. D. Ross (ed.), *The Right and the Good* (Oxford: Clarendon Press, 2002), ix–lviii.
Strawson, P., 'Freedom and resentment', *Proceedings of the British Academy*, 48 (1960).
Suk, J., 'From antidiscrimination to equality: stereotypes and the life cycle in the United States and Europe', *American Journal of Comparative Law*, 60 (2012), 75–98.
Tan, K. C., 'Justice and personal pursuits', *The Journal of Philosophy*, 101/7 (2004), 331–62.
  Justice, Institutions and Luck (Oxford University Press, 2012).
Temkin, L. S., *Inequality* (Oxford University Press, 1993).
  'Justice and equality: some questions about scope', *Social Philosophy and Policy*, 12 (1995), 72–104.
Thomson, J., 'More on the metaphysics of harm', *Philosophy and Phenomenological Research*, 82 (2011), 436–58.
  'Some ruminations on rights', *Arizona Law Review*, 19 (1977), 45–60.
Tobin, J., 'On limiting the domain of inequality', *Journal of Law and Economics*, 13 (1970), 263–77.
The Economist, 'American public opinion and climate change: no green tea', *The Economist Online*, 8 September 2011, www.economist.com/blogs/dailychart/2011/09/american-public-opinion-and-climate-change, last accessed October 10, 2014.
  'Stopping a scorcher', 23 November 2013, www.economist.com/news/books-and-arts/21590347-controversy-over-manipulating-climate-change-stopping-scorcher, last accessed 24 May 2014.
Trail Smelter, Arbitral Tribunal (1939).
Trillionth tonne, www.trillionthtonne.org, last accessed 10 October 2014.
Tronto, J., 'The "Nanny" question in feminism', *Hypatia*, 17 (2002), 34–51.
Tuck, R., *Free Riding* (Cambridge, MA: Harvard University Press, 2009).
Vanderheiden, S., *Atmospheric Justice* (Oxford University Press, 2008).
Victor, D., *Global Warming Gridlock: Creating More Effective Strategies for Protecting the Planet* (Cambridge University Press, 2011).
Von Neumann, J. and O. Morgenstern, *Theory of Games and Economic Behavior* (Princeton University Press, 1944).
Waldron, J., 'Moments of carelessness and massive loss', in D. G. Owen (ed.), *The Philosophical Foundations of Tort Law* (Oxford: Clarendon Press, 1995), pp. 387–40.
Walzer, M., *Spheres of Justice* (Oxford: Basil Blackwell, 1983).
Watson, G., *Agency and Answerability* (Oxford: Clarendon Press, 2004).
Weitzman, M., 'A review of the *Stern Review* on the economics of climate change', *Journal of Economic Literature*, 45 (2007), 703–24.
  'On modeling and interpreting the economics of catastrophic climate change', *Review of Economics and Statistics*, 91/1 (2009), 1–19.

Whyte, K. P., 'Now this! Indigenous sovereignty, political obliviousness and governance models for SRM research', *Ethics, Policy and Environment*, 15 (2012), 172–87.
Winter, G., 'Climate engineering and international law: last resort or the end of humanity?', *Review of European Community and International Environmental Law*, 20/3 (2011), 277–89.
World Bank, http://data.worldbank.org/indicator/EN.ATM.CO2E.PC, last accessed 27 October 2014.
World Health Organisation (WHO), *Global Health Risks: Mortality and Burden of Disease Attributable to Selected Major Risks* (Geneva: WHO Press, 2009).
Wright, R. W., 'Causation in tort law', *California Law Review*, 73 (1985), 1737–828.
Ypi, L., 'Territorial rights and exclusion', *Philosophy Compass*, 8/3 (2013), 241–53.

Index

actions
  collective. *See* collective actions
  expected harms and benefits from. *See* expected harms and benefits
  wrongness and. *See* wrongness
adaptation, 2, 5, 17
  costs, 6, 11
  definition, 41
alternative approach to overdetermination
  levels of description, 176
  probability of being member of actual set, 177–8
  relevance of numbers of contributors, 178–80
  stringency and, 180–2
anthropogenic climate change. *See* climate change
assessment of actions
  objectivating attitude, 206, 208, 212, 213, 214
  performative attitude, 207, 208, 211, 212, 214, 216
  reactive attitude, 151, 206–7, 208, 209, 211, 213, 216

bad actions, emotions following, 206
beneficence, 142, 146, 156
  compliance condition, 137–9, 140
  moral requirement of, 157
burden allocation, 2–3, 6
  domestic justice and, 130
  emissions restrictions and, 139
  historical emissions. *See* historical emissions
  Miller's principle of equal sacrifice and. *See* Miller's principle of equal sacrifice
  polluter pays principle and. *See* polluter pays principle

carbon budget, 9, 73–4, 77
  debate, 2
  division of, 74–5
  emissions exports and, 84–5
  global, spending of, 74
carbon neutral living, arguments for, 156–8
carbon offsetting, 85–6
  project based, 97
carbon sinks. *See also* terrestrial sinks
  attachment arguments for ownership of, 9, 62–4
  capacity, 59
  credit for keeping, 9, 64, 67–8
  distribution of capacity. *See* distributive principles
  improvement arguments for ownership of, 9, 66–8
  ownership of, 61–2
  reservations relating to distributive principles, 60–1, 62
  self-determination arguments for ownership of, 9, 64–6
  sharing costs of, 69–72
carbon trading, 10–11
  advantages of, 10
  cap-allocate-and-trade scheme, 90–2
  objections to, 10–11
causation, 80–1, 174, 175, 214–15
climate change, 217, 219–20
  adaptation. *See* adaption
  agents, 13, 14
  approbation and dissappobation, 206–9
  belief formulation, 14, 224–5
  categories of response to, 44–5
  chance of catastrophe from, 195–7
  distributive justice and, 110–11
  economics, ethics and, 187
  environmentalism and, 221–2
  failure to justify, 216–18
  intragenerational distribution, 109
  issues of life and death and, 189–91
  management, DICE assessment model, 192–3
  mechanisms of, 40–1

249

climate change (cont.)
  mediated responsibility for, 149
  moral challenge of, 56–7
  moral philosophy and, 186, 199
climate change harms, 14, 186
  all things considered reasons, 76, 78–81
  apportioning consequences, 84–7
  compensation for, 86–7, 182–3
  deaths, 189–90
  exported emissions, 73, 85, 86, 88
  fossil fuel exports. *See* fossil fuel exports
  isolation argument and, 79
  'mere causation' and, 80–1
  non-human nature. *See* non-human nature
  offset argument and, 79–80
  overdeterminers of. *See* overdetermination
  population. *See* population
  responsibility for, 81–4
  significant interests of others and, 77, 86
climate change scepticism, 15–16, 220–1
  assessment of, 222–3
  choice of experts problem and, 230–1
  need for progress on, 150
  removal of ideology surrounding, 100–2
  triumph of science over, 231–3
  trust in testimony of others and, 224–5
  understanding, 228–9
  winning of ideological war over, 233
climate justice, 1
  allocation of burdens. *See* burden allocation
  compliance condition and, 138
  equal per capita distribution and, 60
  governmental action and, 145
  individual duties and responsibilities and. *See* individual duties and responsibilities
  luck egalitarianism and, 113–14
  non-institutional duties in relation to, 137, 143, 146
  problem of, 218
  purpose of. *See* purpose of climate justice
  territorial rights and, 61
climate mitigation. *See* mitigation
climate regime, 65
  cap-allocate-and-trade scheme, 90–2
  fairness and, 91
  generalist distributive justice and, 93–4
  geoengineering and, 49
  United Nations, 21–2
club method of treaty construction, 8, 19–21
  assessment of legitimacy of, 35–8
  initial exclusivity in, 33
  intentional costs on non-members, 31–2
  objections to illegitimacy claims, 26–7
  path dependence and unfairness in, 34–5
  pressure on recalcitrant states, 33
  stages of agreement for, 27–8
  unfairness in process, 31
collective action, 138, 144, 149, 155, 157–8, 159, 160, 205, *See also* doing harm
  global, 163
  negligent, 161
  of harming, 157
communicative freedom, nature and, 209–13
communicative interaction, 213–16
communicative reason, 213, 216
costs and benefits, 29
  climate change and, 190
  distribution of, 8, 52, 53, 55
  environmental policies, 54
  expectations of, 187
  governance and, 40

dangerousness
  ethical aspects of, 185
  notion of, 185
  values and, 185
distributive justice, 60, 92, 93
  climate change and, 110–11
  generalist, 93–4
  intragenerational, 113
  Miller's principle of equal sacrifice. *See* Miller's principle of equal sacrifice
  polluter pays principle. *See* polluter pays principle
  scope of principles of, 118
distributive principles
  ability-based, 2, 4, 5–6
  beneficiary principle, 2, 3–5
  equal per capita. *See* equal per capita distribution
  equality principle, 2
  fault-based, 2, 4, *See also* polluter pays principle
  reservations in relation to carbon sinks, 60–1, 62
doing harm, 167–9
  complete causal process and, 168
  constraints against, 168
  normative characteristics of, 168–9
  relevant action, 168

emissions
  caps, fairness and, 90–2, 96, 105, 130
  carbon dioxide equivalent, 73–4, 85
  collective and indirect nature of, 143
  direct, 140, 143
  export of industries which produce, 77–8
  exported, 73, 85, 86, 88
  global justice and, 129–30
  historical. *See* historical emissions

indirect, 140–1, 143, 144
individual. *See* participatory derivation
individual responsibility for. *See* individual duties and responsibilities
just per capita, 139, 140
luxury, 182
national inventories of, 74–5
permissable, allocation of. *See* allocation of permissable emissions
permissable, ethics in allocation of, 184
reduction, 10, 74, 129, 143–4, 148
regulatory reduction framework, 130–1
restrictions, assignment of, 139
tradablity. *See* tradable quotas
value judgements in calculation of. *See* values in emissions calculation
wrongness of, 10, 98
environmentalism, 221–2
equal per capita distribution, 6–7, 9, 59–60, 92–3
climate justice and, 60
reservations relating to sequestering capacity, 60–1, 62
expected harms and benefits, 149, 152–4

fairness, 2, 25, 94, 154, 157, 184
derivative, 156
distributive, 93
emissions caps and, 90–2, 96, 105, 106
intergenerational, 148
public goods and, 70
tradable quotas and, 106
fossil fuel exports, 76–7
acceptance of harm by harmed, 81–2
availability of feasible alternatives to, 77, 83–4
duty to stop external harms from, 86
isolation argument and, 79
offset argument, 79–80
responsibility for harm by exporters, 82, 84
free-riding, 12, 17, 25, 32, 33, 120, 149, 154–5, 156
free will, 206, 214, 216

General Agreement on Tariffs and Trade, 8, 19–20, 34
geoengineering, 8–9
agreement on equitable rules for, 51
carbon dioxide removal, 42–3, 68
cirrus thinning, 43, 44, 45
definition, 39, 42–6
difficulties in categorisation of, 43–4
ethical and political issues, 46–9
ethics of techniques of, 45

governance, 39–40, 49–51
governing institutions, design of, 50–1
greenhouse gas removal, 43, 45
international regulatory regime, 49
ocean fertilisation, 8, 42, 45, 68
order of governance and research, 54–7
research, 39–40, 46, 47–8, 50, 57–8
solar radiation management, 8, 42, 43, 45
techniques with uncertain,
unevenly distributed global effects, 45–7
global cap on emissions
fairness and, 90–2
Pigovian taxation and, 94
global egalitarianism, 119–20
global justice, 46, 71, 92
design of governing insitutions and, 50
emissions and, 129–30
emissions caps and, 130
tradable quotas and, 91
global warming, 11, 24, 41, 73–4, 107, 129, 185
estimates, 41–2
intergenerational justice and, 108–9
luck egalitarian injustice and, 117–18
space dust hypothesis and scepticism and. *See* space dust
unjust inequality and, 109
greenhouse gas accumulation, 20, 232
burning of fossil fuels and, 41
deforestation and, 41
greenhouse gas emissions. *See* emissions
group wrongdoing, 12–13
evaluation of individuals involved in. *See* participatory derivation
without individual wrongdoing, 149

harmful conduct. *See* doing harm
historical emissions
'broken transmissions' and, 3
definition, 2
'exonerating circumstances' and, 3
'legitimate repudiation' and, 3
states and, 2–3

individual duties and responsibilities, 11–13
changes to lifestyle, 12
institutional vigilance, 131–2
justification of, 15
negative, 77
options in absence of just institutions. *See* options in absence of just institutions
overdetermination cases. *See* overdetermination
positive, 77

252　Index

institutional approach to justice, 129, 131–2,
　　See also options in absence of just
　　institutions
　value pluralism and, 135–6
institutional duty
　exclusive reading, 142
　extreme version, 142
　inclusive reading, 142–3
intergenerational justice, 11, 92
　global warming and, 108–9
　luck egalitarianism and. *See* luck
　　egalitarianism
Intergovernmental Panel on Climate Change,
　　39, 41, 44, 75, 185, 224
　guidelines, 75
international community
　morally mandatory aims of, 23–5
　power holders and subjects, 27
International Energy Agency, 82
international environmental law
　legitimacy and illegitimacy in, 28–31
intuition of neutrality, 14, 195, 196–7, 199
　rejection of, 197–8
　strong, 198
　weak, 198

justice. *See* intergenerational justice
　climate. *See* climate justice
　distributive. *See* distributive justice
　domestic, 130
　global. *See* global justice
　institutional approach to. *See* institutional
　　approach to justice
　natural duty of, 142
　theories, distributive principles. *See* distri-
　　butive principles

Kyoto Protocol, 17, 18, 19

luck egalitarian injustice
　generational responsibility and, 112–13
　inequality generating global warming and,
　　117–18
　intergenerational, reduced growth scenario
　　and, 111–16
luck egalitarianism, 11, 108, 109–10, 118–19
　climate justice and, 113–14
　generations and, 121–3
　man-made climate change and, 111
　Miller's principle of equal sacrifice and, 125–7
　national self-determination and, 119–20
　polluter pays principle and, 123–5
　responsibility and, 120–3

Miller's principle of equal sacrifice, 110, 125–7
　luck egalitarianism and, 125–7

mitigation, 1, 7, 11, 17, 21
　costs, 2, 5–6
　definition, 41
　geoengineering. *See* geoengineering
Montreal Protocol, 32
moral considerability
　deontological approach to, 203
　moral status and, 202–6
　non-human nature and. *See* non-human
　　nature
moral phenomena, 207

NESS condition, 174–6
non-human nature, 14–15
　moral considerability and, 203–4
　moral considerability of, 202–6
　obligations to, 204
　reparation, argument from, 204–5

objections to tradable quotas
　actors paid for what they should do, 100, 106
　do it yourself objection, 102–4, 106
　implied right to do wrong, 98–100, 105
　moral stigma removal, 100–2, 106
　unfairly cheap selling, 104–5
options in absence of just institutions
　do nothing, 12, 133, 134–5
　do what they would require, 12, 137–41
　help create just institutions, 12, 134, 142–7
　indeterminacy and, 139–40, 141
　institutional duty and. *See* institutional duty
　institutionalists and, 132–3
　personally promote justice, 12, 134, 135–7
overdetermination, 13, 166–7
　alternative approach to. *See* alternative
　　approach to overdetermination
　denial of existence of, 176
　moral status of, 165–6
　necessary elements of a sufficient set and,
　　174–6
overdetermination-based constraints, 167, 171,
　　175
　justifications for, 172, 174
　sceptics and, 172
　significance of, 177, 178, 181
overdetermining harm, 169–71
　absolutism and, 172–3
　scepticism and, 171–2, 180
　universalisation requirement, 173–4

participatory derivation, 12–13
　difference-making and, 148–50
　international applications, 163–4
　joining in what we should be doing and,
　　158–62
　negative, 12, 150, 159, 160, 163

not joining in what we should be doing, 154–8
positive, 12, 149–50, 154–5, 157, 163
wrongness and, 149
performative attitude, 213
Pigovian taxation, 89, 94
polluter pays principle, 2, 110
  luck egalitarianism and, 123–5
population
  chance of catastrophe from, 195–7
  intuition of neutrality, 199
  intuition of neutrality and. *See* intuition of neutrality
  temperal well-being of, 193
  total utilitarianism and, 193, 194, 199
population ethics, 191–5
  climate change modelling and, 14
  climate change policy and, 199
  intuition of neutrality. *See* intuition of neutrality
  purpose of climate justice, 92–5
  generalist distributive approach, 92, 93–4
  isolationist corrective approach, 92
  isolationist distributive approach, 92
  tradable quotas and, 94–5

rationality, 197, 215–16, 223

scepticism, 171–2
  about climate change. *See* climate change scepticism
sinks
  carbon. *See* carbon sinks
  terrestrial. *See* terrestrial sinks
solar radiation management, 8, 45, 46
  effects of, 47
space dust, 225–6
  confronting scepticism of theory of, 227–8
  hypothesis, 225–6
  scepticism, 226–7
state consent
  imposition of trade restrictions and, 33–4
  justification for constraints on, 25
  modifications to doctrine of, 22–3
  withholding of, 32
state sovereignty, 69
  limitations on, 25, 31
states
  as agents, 163
  fair shares consideration, 82–3
  generations and, 118–20

moral assessment and, 163
negative participatory derivation and, 163
positive participatory derivation and, 163
regulatory institutional ambit of, 136
regulatory schemes, unjust nature of, 12

terrestrial sinks. *See also* carbon sinks
  protection of, 70–2
tradable quotas
  benefits of, 95–7
  coordination and information and, 95–6
  efficiency and, 95
  fairness and, 106
  flexibility and, 95
  global justice and, 91
  objections to. *See* objections to tradable quotas
  project based offsets, 97
  proposed, asylum seekers, 96–7
  purpose of climate justice and, 94–5
  role of, 94
  wrongness of emissions and, 98
treaty construction, 18–21
  club method. *See* club method of treaty construction
  legitimacy and, 21–2
  states and, 8
  universal method, 7–8, 18–19, 37

uncertainty
  normative value of, 51–4
uncertainty, normative value of, 51–4
United Nations Framework Convention on Climate Change, 7, 17, 18–19, 75, 186
  stages of, 14
  stages of international process under, 184
  ultimate objective of, 184
universal method of treaty construction, 37
utilitarianism, 133, 135
  total, 193, 194, 195, 199

values in emissions calculation, 187
  expected, 187–9
  killing, 189–91
values, incommensurabilities of, 190–1

World Health Organization, 189
World Trade Organization, 20
wrongness, 150–1, 152, 154, 162
  non-human nature and, 15
  of emissions, 98

CPSIA information can be obtained
at www.ICGtesting.com
Printed in the USA
LVOW04*1057130116
470464LV00008B/39/P